Teubner Skripten zur
Mathematischen Stochastik

Udo Kamps
A Concept of Generalized
Order Statistics

Teubner Skripten zur Mathematischen Stochastik

Herausgegeben von
Prof. Dr. rer. nat. Jürgen Lehn, Technische Hochschule Darmstadt
Prof. Dr. rer. nat. Norbert Schmitz, Universität Münster
Prof. Dr. phil. nat. Wolfgang Weil, Universität Karlsruhe

Die Texte dieser Reihe wenden sich an fortgeschrittene Studenten, junge Wissenschaftler und Dozenten der Mathematischen Stochastik. Sie dienen einerseits der Orientierung über neue Teilgebiete und ermöglichen die rasche Einarbeitung in neuartige Methoden und Denkweisen; insbesondere werden Überblicke über Gebiete gegeben, für die umfassende Lehrbücher noch ausstehen. Andererseits werden auch klassische Themen unter speziellen Gesichtspunkten behandelt. Ihr Charakter als Skripten, die nicht auf Vollständigkeit bedacht sein müssen, erlaubt es, bei der Stoffauswahl und Darstellung die Lebendigkeit und Originalität von Vorlesungen und Seminaren beizubehalten und so weitergehende Studien anzuregen und zu erleichtern.

A Concept of Generalized Order Statistics

Von Priv.-Doz. Dr. rer. nat. Udo Kamps
Rhein.-Westf. Techn. Hochschule Aachen

Springer Fachmedien Wiesbaden GmbH 1995

Priv.-Doz. Dr. rer. nat. Udo Kamps

Geboren 1959 in Wegberg. Von 1979 bis 1985 Studium der Mathematik und Wirtschaftswissenschaften an der Rheinisch-Westfälischen Technischen Hochschule Aachen, 1985 Diplom in Mathematik, 1987 Promotion und 1992 Habilitation. Von 1990 bis 1994 verschiedene Lehraufträge an der Universität Dortmund und 1993/1994 Lehrstuhlvertretungen an der Christian-Albrechts-Universität zu Kiel.

Die Deutsche Bibliothek – CIP-Einheitsaufnahme

Kamps, Udo:
A concept of generalized order statistics / von Udo Kamps.
(Teubner Skripten zur mathematischen Stochastik)
ISBN 978-3-519-02736-2 ISBN 978-3-663-09196-7 (eBook)
DOI 10.1007/978-3-663-09196-7

Ursprünglich erschienen bei B.G. Teubner Stuttgart 1995

Herstellung: Druckhaus Beltz, Hemsbach/Bergstraße

To my parents

Preface

Order statistics and record values appear in many statistical applications and are widely used in statistical modeling and inference. Both models describe random variables arranged in order of magnitude.

In addition to these well–known models, several other models of ordered random variables, known and new ones, are introduced in this book such as order statistics with non–integral sample size, sequential order statistics, k–th record values, Pfeifer's record model, k_n–records from non–identical distributions and ordered random variables which arise from truncation of distributions.

These models can be effectively applied, e.g., in reliability theory. Here, an order statistic represents the life–length of some r–out–of–n–system which is an important technical structure consisting of n components. For this application, a new and more adequate model is naturally suggested. Sequential order statistics serve as a model describing certain dependencies or interactions among the system components caused by failures of components. Record values are closely connected with the occurrence times of some corresponding non–homogeneous Poisson process and used in so–called shock models. More flexible record models, and therefore more applicable to practical situations, are considered here.

The main purpose of this book is to present a concept of generalized order statistics as a unified approach to a variety of models of ordered random variables.
In the distribution theoretical sense, all of the models mentioned above are contained in the proposed model of generalized order statistics.
Numerous related results on distributional and moment properties of order statistics and k–th record values are found in the literature which are deduced separately; e.g., identities and inequalities for moments, transmission of distributional properties and partial ordering results.
The concept of generalized order statistics, however, enables a common approach to structural similarities and analogies. Well known results can be subsumed, generalized, and integrated within a general framework. In this way, corresponding results for ordinary order statistics and record values can be simultaneously deduced to avoid parallelism and, through integration of known properties, the structure of the embedded models becomes clearer.

Most importantly, the validity of such results is obtained for all types of generalized order statistics and therefore for all the models of ordered random variables mentioned above.

The usefulness of generalized order statistics as a unified approach becomes evident in the first chapter on distribution theory. In the following chapters its use is demonstrated for selected topics such as moments of generalized order statistics, recurrence relations and inequalities for moments and reliability properties.

The concept of generalized order statistics provides a large class of models with many interesting and useful properties for both the description and analysis of practical problems.

The present book is based on my 'Habilitationsschrift' at the Aachen University of Technology. I wish to express my appreciation and gratitude to Professor B. Rauhut, Professor U. Gather and Professor H.A. David for their kind readiness to be reviewers, for their encouragement and valuable comments and suggestions. I would also like to thank the referees of earlier drafts of this material for their constructive criticism which led to a significantly improved presentation. I would like to extend my thanks to the series editors and the publishers for their interest in this book and especially to Professor N. Schmitz for his helpful remarks and support.

Aachen, October 1994 Udo Kamps

Contents

Introduction

Order statistics and record values are widely used in statistical models and inference; both describe random variables arranged in order of magnitude. Hence, the joint distribution of n order statistics or of n record values is defined on some cone of the n–dimensional Euclidean space.

However, there are several other models of ordered random variables with different interpretations and interesting applications, for example, in reliability theory.

Here, a form of the joint density of n ordered random variables is presented that includes all of these structures.

Models of Ordered Random Variables

Order statistics appear in many parts of statistics and they have been extensively investigated. Starting with n random variables, which are usually assumed to be independent and identically distributed, they are arranged in ascending order of magnitude and then called order statistics. So their use in a statistical model of an experiment is obvious, if the realizations already arise in non–decreasing order, for instance, as times of failure of technical components or systems.

A very interesting application of order statistics is found in reliability theory.

The r–th order statistic in a sample of size n represents the life–length of a (n–r+1)–out–of–n–system which is an important technical structure. It consists of n components of the same kind with independent and identically distributed life–lengths. All n components start working simultaneously and the system works, if at least n–r+1 components function; i.e. the system fails, if r or more components fail.

When considering certain aging properties of life–length distributions, results can be found on the transmission of such a distributional property. For example, the increasing failure rate property of the distribution of the components is transmitted to the life–length distribution of any (n–r+1)–out–of–n–system.

A more flexible and more adequate model for a (n–r+1)–out–of–n–system has to take a specific dependence structure into consideration. If some component of the system fails, this

may have an influence on the life–length distributions of the remaining components.
Thus, a modification of order statistics is naturally suggested.
Consider n components with life–length distribution F_1. At time x we observe the first failure. Then we suppose that the remaining components possess the life–length distribution F_2 truncated on the left at x, etc.
Proceeding in this way, we obtain a general structure which we call the model of sequential order statistics. It coincides with ordinary order statistics in the case of identical underlying distribution functions $F_1,...,F_n$.
Here, we restrict ourselves to a particular choice of these distribution functions.
Some questions arise. Is it possible to obtain the distribution theory of sequential order statistics by analogy with ordinary order statistics? There is a variety of well–known and useful properties of order statistics. Do we have analogous results for the enlarged model? In particular we may ask whether we have transmission results for aging properties when analyzing modified (n–r+1)–out–of–n–systems.

Order statistics with non–integral sample size have been introduced as an extension of ordinary order statistics. In order to notice a practical application, these quantities can be interpreted as certain sequential order statistics.

When considering models of ordered random variables, we are led to several models of record values.
Motivated by extreme weather conditions, record values were defined as a model for successive extremes in a sequence of independent and identically distributed random variables. Record values have been extensively examined and many useful properties are known.
Several applications of this model can be found, for example, in reliability theory.
Suppose that a technical system is subject to shocks, e.g. peaks of voltage. These shocks may be modelled as realizations of records.
Since record values are closely connected with the occurrence times of some corresponding non–homogeneous Poisson process, we have another possibility for their application. Consider a situation in which the times between shocks, and not the values of the shocks, are to be modelled. If we suppose that the shocks appear at the occurrence times of some non–homogeneous Poisson process, then the waiting times between each two shocks can be described by record differences.
Hence, the model of record values can be used, if we are interested in the values of successive peak voltages as well as if we consider the occurrence times of these peaks.
By analogy with the transmission rules for aging properties in the case of order statistics

there are results for record values. For example, if the underlying distribution function has an increasing failure rate, then all distribution functions of the records possess this property.

If not the record values themselves, but second or third largest values are of special interest, then the model of k–th record values is adequate where k is some positive integer. Obviously, putting $k = 1$ we obtain ordinary record values. Moreover, k–th record values based on a distribution function F can be viewed as ordinary record values based on the distribution function of the minimum of k random variables which are independent and identically distributed according to F.

Pfeifer's record model is based on non–identically distributed random variables. The distribution of the underlying random variables may change after each record event.
For example, interventions can be described as modifying the situation after the occurrence of a record. Through this, we obtain a class of models which can be applied in reliability theory. Used as a shock model, each shock is now allowed to influence the magnitude of the subsequent one.
Here, we restrict ourselves to a particular choice of $F_1,...,F_n$ where F_r denotes the underlying distribution function until the r–th record occurs.
Choosing identical distributions, we see that record values are contained in Pfeifer's model in the distribution theoretical sense.
By analogy with sequential order statistics some questions arise. For instance, is it possible to generalize well–known properties of ordinary records? Do we have transmission results for aging properties in Pfeifer's record model?

Furthermore, Pfeifer's model and the model of k–th record values can be combined to obtain an enlarged model which may provide a better fit for practical situations. More generally, given a sequence $k_1,k_2,...$ of positive integers, we are successively looking for the k_1–th largest value (based on F_1), then for the k_2–th largest value (based on F_2), etc. Obviously, we obtain Pfeifer's records by putting $k_n = 1$ for all n. If all distribution functions are identical, then this model may be viewed as a generalization of k–th record values.

Several models of censored data can be described in terms of order statistics, since the observations arise in ascending order of magnitude. In a special model of progressive type II censoring the corresponding joint density function of the quantities coincides with the joint density of certain sequential order statistics as well as of certain record values from non–identical distributions.

Generalized Order Statistics

In the distribution theoretical sense, all of these models of ordered random variables are contained in the model of generalized order statistics which we propose.
The starting point is a joint density function on a cone of the n–dimensional Euclidean space. Random variables possessing such a density are called uniform generalized order statistics. Generalized order statistics based on some distribution function F are then defined by means of the quantile transformation. Assuming an absolutely continuous distribution function F and choosing the parameters in the joint density of generalized order statistics appropriately, we obtain the densities corresponding to the models of ordered random variables mentioned above.

In particular, ordinary order statistics and record values are contained in the model of generalized order statistics which can be viewed as a parametrical generalization.
There are numerous analogies in the properties and the behaviour of order statistics and record values :

For instance, the exponential distribution can be characterized by the independence of functions of order statistics (cf. Fisz 1958, Rossberg 1960, Ferguson 1964, Crawford 1966) as well as by the independence of functions of records (cf. Tata 1969).
If the exponential distribution is the underlying distribution, then successive differences of order statistics and of record values are independent and again exponentially distributed (Sukhatme 1937, Rényi 1953 and Tata 1969, Resnick 1973a).
Most importantly, the models have the Markov property in common with similarly structured transition probabilities.
These relationships are taken up by Deheuvels (1984) and Gupta (1984). Instead of parallel investigations, the common Markov structure leads to a unified approach to results concerning independent and identically distributed statistics and to the asymptotic behaviour; implicitly this is already contained in the dissertation of Pfeifer (1979) (see Pfeifer 1982a,b).
We observe analogies concerning concomitants of order statistics (David 1973, Yang 1977) and concomitants of records (Houchens 1984) and there is a variety of related results on moments of order statistics and record values. We find related sufficient conditions for the existence of moments, and certain sequences of moments of order statistics as well as of record values characterize the underlying distribution (Hoeffding 1953, Hwang, Lin 1984a, Huang 1989, Lin 1989a and Kirmani, Beg 1984, Lin 1987). Sequences of moment differences

of order statistics and records characterize the underlying distribution up to a location parameter (Saleh 1976, Lin 1988b and Gupta 1984).
Recurrence relations for moments of order statistics and record values are similar (Kamps 1992a) and related inequalities between moments exist. When characterizing equality, characterization theorems result (Lin 1988a, Gajek, Gather 1991, Kamps 1991a). For other analogies see, e.g., Kakosyan, Klebanov, Melamed 1984, Aly 1988 and the literature on conditional distributions, conditional moments and asymptotic distributions. We previously mentioned the behaviour of order statistics and record values concerning the transmission of aging properties (Takahasi 1988, Nagaraja 1990, Gupta, Kirmani 1988, Kochar 1990). Occasionally, the cited analogies are somewhat hidden in view of the different numbering of record values.

The proposed concept of generalized order statistics enables a unified approach to the cited structural similarities and analogies. Well–known properties of order statistics and of record values can be subsumed, generalized and integrated within a general framework:

Corresponding results for order statistics and record values, e.g. inequalities for moments, can be deduced at once to avoid parallelism. Moreover, we obtain the validity of such results for generalized order statistics. Hence, these results not only hold true for ordinary order statistics and record values, but also for all models of ordered random variables mentioned above.

In other cases, we have some useful properties of order statistics, e.g. recurrence relations for moments of order statistics. Through generalizing such a relation to generalized order statistics and by choosing the parameters appropriately, we obtain new corresponding recurrence relations, e.g. for moments of sequential order statistics, record values and of Pfeifer's records.
Several examples are presented in the following chapters.

In the consideration of sufficient conditions for the existence of moments of generalized order statistics, we observe different behaviour with respect to the parameters. Since order statistics and record values are two particular examples of generalized order statistics, it becomes clear why different conditions are found in the literature.
In the model of generalized order statistics, the joint density on some cone of the n–dimensional Euclidean space is indexed by several parameters. Order statistics and record values correspond to very special cases. Thus, different distributions arising in characterization results by means of these two sub–models are now elements of a certain

parametrized class of probability distributions and are therefore related.
In this sense, generalized order statistics serve as a model explaining the striking analogies in the behaviour of order statistics and record values as well as certain differences. Considering the several models of ordered random variables, the structure of these embedded models becomes clearer.
On the other hand, we derive new results for different models of ordered random variables. For instance, we show transmission rules with respect to aging properties not only for order statistics and record values, but also for sequential order statistics (to describe modified $(n-r+1)$–out–of–n–systems) and k–records, Pfeifer's records, k_n–records (which can be used in shock models).

It will be shown that many well known properties of ordinary order statistics and record values are also valid for generalized order statistics. Thus, we have a large class of models with useful properties at our disposal to both describe and analyze practical problems.

It should be noted that the term 'generalized order statistics' has been used by Choudhury, Serfling (1988) in a different context.

In the sequel, several characterization results are stated, e.g. by means of moments and inequalities for moments of generalized order statistics. Reflections on the advantages of characterization results can be found in Galambos, Kotz (1978). In order to fit a distribution to a given data set, one usually chooses a parametrized distribution in a certain class of distributions and estimates the parameters. However, the question arises as to whether this procedure is always reasonable. Preferably, preliminary or additional information on distributional properties should be used to fix a distribution or at least a class of distributions by means of a characterization result. For example, the 'lack of memory property' of the exponential distribution justifies its application in reliability or insurance models.

Outline of the Following Chapters

In the first section of Chapter I the models of ordered random variables previously mentioned are described and discussed in some detail. In the distribution theoretical sense, these structures are contained in the model of generalized order statistics which we

introduce in Section 2. After defining uniform generalized order statistics and using the quantile transformation, we then turn to generalized order statistics based on an arbitrary distribution function F.
The following abbreviations will be used:

g OS : generalized order statistic, **o OS** : ordinary order statistic,
OS′s : order statistics.

Section 3 contains the distribution theory for uniform g OS′s and by this for g OS′s based on F. Representations for the one–, two– and higher dimensional marginal density functions are given as well as a form of the one–dimensional marginal distribution functions. The marginal distribution of a single uniform g OS provides the basis for the results obtained in the following chapters. Thus, we show some of its properties as, for instance, recurrence relations for marginal density functions and distribution functions. Transformations within the class of g OS′s and properties based on the Markov property are shown; by this, structural aspects become clear. Finally, concomitants of g OS′s are introduced including concomitants of o OS′s as introduced by David (1973) (see Yang 1977) and those of records (Houchens 1984). Examples of results on g OS′s are stated in terms of the models of ordered random variables shown in Section 1.

In Chapter II we show representations for moments and differences of moments of g OS′s as well as results on the existence of moments used in the second section and in the following two chapters. Moments of g OS′s based on some distribution function F can be determined using the pseudo–inverse of F and the marginal densities of the corresponding uniform g OS′s. Sufficient conditions for the existence of moments of g OS′s are stated next; in particular, Sen′s (1959) well–known theorem is formulated for g OS′s. To prepare for the second section and for recurrence relations for moments, we derive two representations for the difference of moments of successive g OS′s with different conditions imposed on the underlying distribution function. In the case of o OS′s these representations are stated in David, Shu (1978), Khan, Yaqub, Parvez (1983), Lin (1988b). Following, we show explicit expressions for the moments of g OS′s based on power function, Pareto and Weibull distributions. Proceeding from characterizations of distributions by sequences of moments of o OS′s and the well–known work of Hoeffding (1953), the second section deals with characterizations of distributions by means of certain sequences of expectations of corresponding g OS′s. We cite some known results on complete function sequences and then apply them to obtain characterizing sequences of expectations and sequences of differences of expectations of g OS′s. In particular, results on o OS′s (Galambos 1975, Hwang, Lin 1984a, Huang 1989, Lin 1989a, Saleh 1976, Lin 1988b) and records (Kirmani, Beg 1984, Lin 1987, Gupta 1984) are contained.

Numerous articles on recurrence relations for moments of o OS′s are found in the literature; reasons and an introduction are given at the beginning of Chapter III. Identities for records have been considered only incidentally as they appear as a by–product when characterizing equality in inequalities for moments of records. The results up to now may be described as rather isolated; explicit expressions for the moments of some distribution lead to a recurrence relation. A step towards a systematic treatment is shown in Khan, Yaqub, Parvez (1983) and Lin (1988b). They derive a representation for the difference of moments of successive o OS′s and put in special distributions leading to similar recurrence relations. Azlarov, Volodin (1986) and Lin (1988b) state corresponding characterization results assuming the validity of some identity for a certain sequence of order statistics. We observe that there are similarly structured recurrence relations. This is the motivation for an attempt at a unified approach to such identities. We proceed as follows. The starting point is a parametrized recurrence relation. Following, a characterization set–up may lead to a corresponding parametrized family of distributions applying an appropriate complete sequence of functions. Going backwards, the strong assumptions are dropped and, under mild conditions, the relation is verified within this class of distributions (Kamps 1991b, 1990a). This approach provides an insight into structural properties and relationships of several probability distributions. Moreover, isolated results can be subsumed and well known results can be generalized with respect to the parametrization of the underlying distributions and to moments of non–integral orders. This method and its application to g OS′s is demonstrated for three classes of distributions. By this, the integration of known results concerning o OS′s and records becomes clear. We show several examples of distributions contained in the classes considered. The third chapter ends generalizing a characterization of Too, Lin (1989) by means of two moments of g OS′s.

Chapter IV contains inequalities for moments of g OS′s and corresponding characteri–zations of probability distributions. First results are stated by Plackett (1947), Moriguti (1951), Gumbel (1954), Hartley, David (1954) and Nagaraja (1978) for o OS′s and record values characterizing certain power function and Weibull distributions. In contrast to the ones deduced by means of recurrence relations, characterization results are derived under mild conditions and, simultaneously, recurrence relations are obtained for those distributions characterized by equality.
The work of Lin (1988a) in the case of o OS′s and records is taken up by Gajek, Gather (1991), Kamps (1991a) and generalized, e.g. in regard to the order of moments, using Hölder′s inequality and its inverse version. Here, we establish results for g OS′s along these lines covering the ones cited above. Finally, other integral inequalities are applied.

Properties of the hazard rate of g OS′s, e.g. IFR or DFR property, and partial ordering of g OS′s are subject matters of Chapter V. We generalize results of Takahasi (1988), Nagaraja (1990), Gupta, Kirmani (1988) and Kochar (1990) on o OS′s and records to obtain properties of g OS′s concerning the transmission of aging properties of life–length distributions. Moreover, we consider several partial orderings between random variables or distribution functions and show results for g OS′s.

The Appendix contains some integration formulae and we prove useful combinatorial identities.

Chapter I

Generalized Order Statistics

Several models of ordered random variables are discussed in Section 1. In the distribution theoretical sense, these structures are included in the model of generalized order statistics which we introduce next. Section 3 contains the basic distribution theory of generalized order statistics. Representations of one– and higher dimensional marginal density functions and of the distribution function of a single generalized order statistic are derived and useful properties of them are shown.

1. Models of Ordered Random Variables

In addition to the well–known models of order statistics and record values there are several other models of ordered r.v.'s. These were briefly mentioned in the introduction and will be described and discussed in some detail in this section. In particular, we point out new aspects and possibilities of different interpretations.
In the distribution theoretical sense, i.e. choosing the parameters appropriately, all of the following models are contained in the model of generalized order statistics presented in Section 2.

1.1. Order Statistics

Order statistics appear in many parts of statistics and play an important role in statistical modelling.
If the random variables $X_1, X_2, \ldots, X_n$

are arranged in ascending order of magnitude, then the ordered quantities

$$X_{1,n} \leq X_{2,n} \leq \ldots \leq X_{n,n}$$

are called order statistics.

To be more precise, we state two possible definitions of order statistics.

Let $X_1,X_2,\dots,X_n$ be random variables on some probability space.

Definition 1.1.1. Let the function $T : \mathbb{R}^n \longrightarrow \mathbb{R}^n$ be defined by

$$T(x_1,\dots,x_n) = (x_{1,n},\dots,x_{n,n}), \quad (x_1,\dots,x_n) \in \mathbb{R}^n,$$

such that
$$x_{1,n} \leq \dots \leq x_{n,n}$$

and
$$(x_1,\dots,x_n)\,\mathsf{P} = (x_{1,n},\dots,x_{n,n})$$

where P is a $n{\times}n$ permutation matrix (which results from permuting the columns of an $n{\times}n$ identity matrix).

Moreover, let the functions $T_i : \mathbb{R}^n \longrightarrow \mathbb{R}^1$ be defined by

$$T_i(x_1,\dots,x_n) = x_{i,n}, \quad 1 \leq i \leq n .$$

Then, with real valued random variables $X_1,\dots,X_n$,

$$X_{i,n} = T_i(X_1,\dots,X_n)$$

is called the i–th order statistic, $1 \leq i \leq n$, and

$$(X_{1,n},\dots,X_{n,n}) = T(X_1,\dots,X_n)$$

is the vector of order statistics (based on $X_1,\dots,X_n$).

In this definition, the ordering of the initial r.v.'s $X_1,\dots,X_n$ is described via the measurable functions T and T_i where T maps some vector of $\mathbb{R}^n$ to a vector with ordered components.

If the underlying r.v.'s have a discrete distribution, identical values, called ties, may be observed. The condition with the permutation matrix P ensures that, in the presence of ties, the multiplicity of the values is preserved. In the case of continuously distributed r.v.'s, ties occur with probability zero and therefore can be ignored.

In the second definition of order statistics, the pseudo–inverse function F^{-1} of some distribution function F is used which we introduce first.

Definition 1.1.2. Given the distribution function F corresponding to a one-dimensional probability distribution, then the function $F^{-1} : (0,1) \longrightarrow \mathbb{R}$ with

$$F^{-1}(y) = \inf\{x; F(x) \geq y\}, \quad y \in (0,1),$$

is called pseudo-inverse (or quantile function) of F.

Moreover, let $F^{-1}(0) = F^{-1}(0+)$ and $F^{-1}(1) = F^{-1}(1-)$.

By this, the order statistics can be defined in an alternative manner.

Definition 1.1.3. The random variable $X_{i,n}$ defined by

$$X_{i,n} = \hat{F}_n^{-1}\left(\frac{i}{n}\right)$$

is called the i-th order statistic, $1 \leq i \leq n$, where $\hat{F}_n^{-1}$ is the pseudo-inverse of the empirical distribution function $\hat{F}_n$ with

$$\hat{F}_n(x) = \frac{1}{n}\sum_{i=1}^{n} 1_{(-\infty,x]}(X_i), \quad x \in \mathbb{R}.$$

When introducing order statistics, no restrictions are imposed on the underlying r.v.'s. Usually it is assumed that order statistics are based on independent and identically distributed r.v.'s. Otherwise, the distribution theory becomes quite complicated.

Assumption 1.1.4. Throughout it is assumed here that order statistics are based on independent and identically distributed random variables $X_1,\dots,X_n$ with distribution function F.

Concerning distribution theory, properties and statistical applications of order statistics, we refer to the basic monograph of David (1981) as well as to Castillo (1988), Arnold, Balakrishnan (1989), Reiss (1989), Balakrishnan, Cohen (1991) and Arnold, Balakrishnan, Nagaraja (1992).

In David's well-known book, results and detailed discussions are found on distribution theory for order statistics, expected values, bounds and approximations for moments, order statistics in statistical estimation and hypothesis testing, treatment of outliers and on asymptotic theory. Castillo (1988) presents the application of order statistics and extreme value theory in engineering and shows several practical examples. Arnold, Balakrishnan (1989) focus on recurrence relations, bounds and approximations for moments of order statistics. Reiss (1989) concentrates on a detailed account of the asymptotic theory of order statistics and its statistical aspects. Balakrishnan, Cohen (1991) deal with the basic theory of order statistics, relations for moments of order statistics, and they give a survey on estimation based on order statistics with a special emphasis on best linear unbiased estimation and maximum likelihood estimation. The book by Arnold, Balakrishnan, Nagaraja (1992) is an introductory text in order statistics covering distribution theory, moment relations, statistical inference and asymptotic theory.

In David (1981), Reiss (1989), Balakrishnan, Cohen (1991) and Arnold, Balakrishnan, Nagaraja (1992), brief descriptions of the developments in some areas of the theory of order statistics are found. A detailed survey on the history of order statistics is given by Harter (1988).

The use of order statistics in a statistical model of an experiment is obvious, if the realizations, e.g. the measurements, arise in order of magnitude as, e.g., in the case of times of failure or life–lengths of technical systems and censored data.

Moreover, order statistics are needed in the description of outliers; see Barnett, Lewis (1978, 1994), Gather (1984), Barnett (1988), Gather, Rauhut (1990) and the cited literature therein. On the other hand, the information contained in extremal observations is often of interest. A large number of publications deals with extreme value theory and its applications. We refer to the books of Galambos (1978), Leadbetter, Lindgren, Rootzén (1983), Resnick (1987), Castillo (1988), Pfeifer (1989) and Reiss (1989). Other fields of application of order statistics are, e.g., robust procedures (cf. Huber 1981) and goodness of fit tests (Tiku 1988, Padwardhan 1988).

1.1.5. Based on iid r.v.'s $X_1,\dots,X_n$ possessing an absolutely continuous distribution function F and density function f, we obtain the joint density of the corresponding order statistics $X_{1,n},\dots,X_{r,n}$

$$f^{X_{1,n},\dots,X_{r,n}}(x_1,\dots,x_r) = \frac{n!}{(n-r)!}\left(\prod_{i=1}^{r-1} f(x_i)\right)(1-F(x_r))^{n-r} f(x_r)\,,\quad r \le n\,,$$

the marginal densities

$$f^{X_{r,n}}(x) = r\binom{n}{r} F^{r-1}(x)\,(1-F(x))^{n-r} f(x)\,,\quad 1 \le r \le n\,,$$

and the marginal distribution functions

$$F^{X_{r,n}}(x) = 1-(1-F(x))^{n-r+1} \sum_{j=0}^{r-1} \binom{n-r+j}{j} F^j(x)\,, \quad 1 \leq r \leq n\,.$$

Moreover, the order statistics form a Markov chain with transition probabilities

$$P(X_{r,n} > t \mid X_{r-1,n} = s) = \left(\frac{1-F(t)}{1-F(s)}\right)^{n-r+1}, \quad 2 \leq r \leq n\,.$$

The Markov property of order statistics from a continuous distribution is well known (e.g. David 1981, p 20). Apart from the continuity of F, Nagaraja (1982) shows that the order statistics in a sample of size $n \geq 3$ from a discrete distribution form a Markov chain iff its support consists of one or two points. Arnold, Becker, Gather, Zahedi (1984) present a necessary and sufficient condition for the order statistics ($n \geq 3$) to have a Markovian structure; i.e. there does not exist any atom x_0 of the parent distribution F such that $F(x_0-) > 0$ and $F(x_0) < 1$. By considering the multiplicities of ties, Rüschendorf (1985) shows a certain Markov property of order statistics from an arbitrary distribution.

The Markovian structure of order statistics turns out to be a very important tool in order to derive several distributional properties and characterization results. E.g., it is applied in Pyke (1965) for determining the joint distribution of spacings and in Galambos, Kotz (1978) for proving characterization theorems.

Another very interesting application of order statistics is found in reliability theory. The r–th order statistic $X_{r,n}$ in a sample of size n represents the life–length of a (n–r+1)–out–of–n–system which is a well known and important technical structure (e.g. Barlow, Proschan 1975, David 1981).
This system consists of n components of the same kind with independent and identically distributed life–lengths. All n components start working simultaneously and the system fails, if r or more fail. In other words, n–r+1 components are necessary for the system to work. (Failed components are supposed to be exchanged without any loss of time.)
For $r = 1$ we have a series system and the case $r = n$ corresponds to a parallel system.

When considering certain aging properties of life–length distributions, there are results on the transmission of such a distributional property. E.g., the IFR property (increasing failure rate) is transmitted from the distribution of the components to the life–length distribution of some (n–r+1)–out–of–n–system (Barlow, Proschan 1965). This result is taken up by Takahasi (1988) and extended to the transmission of the IFR property from some order statistic to the next one with respect to the parameter r. That is, the IFR property of the life–length distribution of some (n–r+1)–out–of–n–system ensures the IFR property of a neighbouring (n–r)–out–of–n–system. Moreover, Nagaraja (1990) shows transmission results with respect to other neighbouring order statistics or systems. For more details we refer to Chapter V.

A certain modified (n–r+1)–out–of–n–system in connection with the possibility of a more flexible and more adequate statistical model for such structures is the motivation for defining the model of sequential order statistics presented in Section 1.3.

In the sequel, order statistics are often called ordinary order statistics to differentiate them from other models.

1.2. Order Statistics with Non–Integral Sample Size

As an extension of the model of order statistics, Stigler (1977) defines fractional order statistics by means of some order statistics process. These quantities coincide with ordinary order statistics $X_{r,n}$, if a positive integer valued parameter r is chosen.

Motivated by this work, Rohatgi, Saleh (1988) start with the distribution function $F^{X_{r,n}}$ of the order statistic $X_{r,n}$ and extend this to the case when the sample size $n = \alpha$ is a positive real number.

Thus
$$F^{X_{r,\alpha}}(x) = r\binom{\alpha}{r}\int_0^{F(x)} t^{r-1}(1-t)^{\alpha-r}\,dt\,,$$

$$r\binom{\alpha}{r} = \frac{\alpha\,(\alpha-1)\cdot\ldots\cdot(\alpha-r+1)}{(r-1)!}\,,$$

for some r and α satisfying $0 < r < \alpha$, which coincides with $F^{X_{r,n}}$ in the previous section, if $\alpha = n$ is a natural number.

The idea is to use binomial series expansion of $(F + (1 - F))^{\alpha}$ with an arbitrary distribution function F.

The same argument leads to a representation of some joint distribution of suitable r.v.'s

$$X_{r_1,\alpha}, \dots, X_{r_k,\alpha} \text{ with } 1 \le r_1 \le \dots \le r_k \le \alpha, \; r_1,\dots,r_k \in \mathbb{N},$$

which are viewed as order statistics with non-integral sample size.

But no practical interpretations or applications are noted. Thus, for the present, this model is of theoretical value.
In Section 2 however, we will see that order statistics with non-integral sample size can in fact be interpreted, namely as sequential order statistics or as certain records.

Choosing $r_i = i$, $1 \le i \le n$, $\alpha = n$ for some $n \in \mathbb{N}$, and assuming F to be absolutely continuous with density f, the corresponding joint density function coincides with the joint density of ordinary order statistics based on the distribution function F.

1.3. Sequential Order Statistics

In this section we introduce a model of ordered r.v.'s which, in some sense, extends the ordinary model of order statistics.
This modification of order statistics is suggested by a statistical application in reliability theory.
Recalling (n–r+1)–out–of–n–systems (see end of Section 1.1.), it is generally assumed that we have components of the same kind without any interactions with respect to life–length distributions. Hence, the system failure is modelled by an order statistic based on iid r.v.'s.

However, the failure of some component can more or less strongly influence the remaining components. This can be thought of as damage caused by the i–th failure in the system.
Thus, a more flexible model, that is more general and therefore more applicable to practical situations, must take some dependence structure into account.

In the following model, the life–length distribution of the remaining components in the system may change after each failure of the components. If we observe the i–th failure at time x, the remaining components are now supposed to have a possibly different life–length distribution. This distribution is truncated on the left at x to ensure realizations arranged in ascending order of magnitude.

DEFINITION 1.3.1. Let $(Y_j^{(i)})_{1\le i\le n,\ 1\le j\le n-i+1}$ be independent r.v.'s

with $$(Y_j^{(i)})_{1\le j\le n-i+1} \sim F_i\,,\quad 1\le i\le n\,,$$

where $F_1, \dots, F_n$ are strictly increasing and continuous distribution functions

with $$F_1^{-1}(1) \le \dots \le F_n^{-1}(1)\,.$$

Moreover, let $$X_j^{(1)} = Y_j^{(1)}\,,\quad 1\le j\le n\,,$$

$$X_*^{(1)} = \min\{\, X_1^{(1)},\dots,X_n^{(1)} \,\}$$

and for $2\le i\le n$:

$$X_j^{(i)} = F_i^{-1}\Big(F_i(Y_j^{(i)})\,(1-F_i(X_*^{(i-1)})) + F_i(X_*^{(i-1)}) \Big)\,,$$

$$X_*^{(i)} = \min\{\, X_j^{(i)},\ 1\le j\le n-i+1 \,\}\,.$$

Then the r.v.'s $$X_*^{(1)}, \dots, X_*^{(n)}$$

are called **sequential order statistics.**

In this definition we start with some triangular scheme of r.v.'s, the i–th line containing $n-i+1$ r.v.'s with distribution function F_i, $1\le i\le n$.
The condition on the supports of $F_1,\dots,F_n$ ensures that truncation of F_i on the left at x is possible for any realization x of $Y_1^{(i-1)}$.
Putting $X_*^{(0)} = -\infty$, the definition of $X_j^{(i)}$ for $2\le i\le n$ also holds for $i = 1$.

Given the realization of the minimum in line $i-1$, we obtain the conditional distribution of the r.v.'s $X_1^{(i)}$ for $1\le i\le n$.

1.3.2. Since $Y_1^{(i)}$ and $X_*^{(i-1)}$ are independent, we have

$$P(\, X_1^{(i)} \le t \mid X_*^{(i-1)} = s\,) \;=\; P(\, F_i(Y_1^{(i)}) \le \frac{F_i(t) - F_i(s)}{1 - F_i(s)}\,) \;=\; \frac{F_i(t) - F_i(s)}{1 - F_i(s)}\,.$$

In view of 1.3.2. we describe our modified $(n-r+1)$–out–of–n–system as follows :

After the occurrence of the i–th failure in the system at time $z_{1,n-i+1}^{(i)}$ (i.e. the realization of the sample minimum in line i), the next failure time is modelled as the minimum in the sample

$$Z_1^{(i+1)},\dots,Z_{n-i}^{(i+1)}$$

of iid r.v.'s with distribution function

$$\frac{F_{i+1}(\cdot) - F_{i+1}(z_{1,n-i+1}^{(i)})}{1 - F_{i+1}(z_{1,n-i+1}^{(i)})}.$$

Thus, we may illustrate the model of sequential order statistics in this way:

Consider a triangular scheme $(Z_j^{(i)})_{1\leq i\leq n,\ 1\leq j\leq n-i+1}$

of r.v.'s where the $(Z_j^{(i)})_{1\leq j\leq n-i+1}$

are iid according to

$$\frac{F_i(\cdot) - F_i(z_{1,n-i+2}^{(i-1)})}{1 - F_i(z_{1,n-i+2}^{(i-1)})},\quad 1\leq i\leq n,\ z_{1,n+1}^{(0)} = -\infty .$$

$$\begin{array}{ccc|ccccc|l}
X_*^{(1)} & \leftarrow & & Z_1^{(1)} & Z_2^{(1)} & \dots & Z_{n-1}^{(1)} & Z_n^{(1)} & \sim F_1(\cdot) \\
X_*^{(2)} & \leftarrow & & Z_1^{(2)} & Z_2^{(2)} & \dots & Z_{n-1}^{(2)} & & \sim \dfrac{F_2(\cdot) - F_2(z_{1,n}^{(1)})}{1 - F_2(z_{1,n}^{(1)})} \\
\cdot & & & \cdot & \cdot & \cdot & & & \quad\cdot \\
\cdot & & & \cdot & \cdot & \cdot & & & \quad\cdot \\
\cdot & & & \cdot & \cdot & \cdot & & & \quad\cdot \\
X_*^{(n-1)} & \leftarrow & & Z_1^{(n-1)} & Z_2^{(n-1)} & & & & \\
X_*^{(n)} & \leftarrow & & Z_1^{(n)} & & & & & \sim \dfrac{F_n(\cdot) - F_n(z_{1,2}^{(n-1)})}{1 - F_n(z_{1,2}^{(n-1)})} \\
\uparrow & & & & & & & &
\end{array}$$

line minima $X_*^{(1)} \leq \dots \leq X_*^{(n)}$

1.3.3. If we have absolutely continuous distribution functions $F_1,...,F_n$ with densities $f_1,...,f_n$ respectively, then it can be shown that the joint density of the first r sequential order statistics $X_*^{(1)},...,X_*^{(r)}$ is given by

$$f^{X_*^{(1)},...,X_*^{(r)}}(x_1,...,x_r) = \frac{n!}{(n-r)!}\left(\prod_{i=1}^{r-1}\left(\frac{1-F_i(x_i)}{1-F_{i+1}(x_i)}\right)^{n-i} f_i(x_i)\right)(1-F_r(x_r))^{n-r} f_r(x_r)$$

$$= \frac{n!}{(n-r)!}\prod_{i=1}^{r}\left(\frac{1-F_i(x_i)}{1-F_i(x_{i-1})}\right)^{n-i}\frac{f_i(x_i)}{1-F_i(x_{i-1})}, \quad r \le n, \quad x_0 = -\infty .$$

Moreover, sequential order statistics form a Markov chain with transition probabilities

$$P(X_*^{(r)} > t \mid X_*^{(r-1)} = s) = \left(\frac{1-F_r(t)}{1-F_r(s)}\right)^{n-r+1}, \quad 2 \le r \le n .$$

Ordinary order statistics are contained in this model in the distribution theoretical sense. Choosing $F_1 = ... = F_n$ ($= F$), we obtain the joint density function of the order statistics $X_{1,n},...,X_{n,n}$ based on n iid r.v.'s with distribution function F (see 1.1.5.).

At this point the question arises as to whether we can obtain the distribution theory of sequential order statistics and their properties by analogy with ordinary order statistics. There is a variety of useful properties of order statistics. Do we have analogous results for the enlarged model?

Sequential order statistics are motivated by some modified (n–r+1)–out–of–n–system. Do we find transmission results for aging properties by analogy with the results in Section 1.1.?

However, in the general setting the model of sequential order statistics turns out to be too extensive. E.g., there is no simple representation of a marginal density which is fundamental for many applications.

In the distribution theoretical sense, this model is as general as Pfeifer's record model based on non–identically distributed random variables. We will return to the underlying Markovian structure in Section 1.6.

If we aim at establishing analogous properties as found in the case of ordinary order statistics, we have to look for a suitable subclass within the general model.

1.3.4. In the following we restrict ourselves to a particular choice of the distribution functions $F_1,...,F_n$, namely

$$F_r(t) = 1-(1-F(t))^{\alpha_r}, \quad 1 \leq r \leq n,$$

with some distribution function F and positive real numbers $\alpha_1,...,\alpha_n$ (cf. 1.6.8. in the case of Pfeifers's records).

It will be shown in the sequel that a variety of well–known properties of order statistics can also be stated in terms of sequential order statistics.
As a result, we now have a wide class of models at hand to both describe and analyze modified (n–r+1)–out–of–n–systems; i.e. systems in which certain dependencies or interactions between the components (as described above) can be taken into consideration.

1.4. Record Values

Record values are found in many situations of daily life as well as in many statistical applications. Often we are interested in observing new records and in recording them: e.g. Olympic records or world records in sports.

Motivated by extreme weather conditions, record values are defined by Chandler (1952) as a model for successive extremes in a sequence of iid r.v.'s. It may also be helpful as a model for successively largest insurance claims in non–life insurance, for highest water–levels or highest temperatures. Record values are also used in reliability theory.
Suppose that a technical system or a piece of equipment is subject to shocks, e.g. peaks of voltage. If the shocks are viewed as realizations of an iid sequence, then the model of record values is adequate. Another interpretation in connection with a shock model is presented below.

To be precise, record values are defined by means of record times. That is, those times have to be described at which successively largest values appear.

DEFINITION 1.4.1. Let $(X_i)_{i\in\mathbb{N}}$ be a sequence of iid r.v.'s with a continuous distribution function F. The r.v.'s

$$L(1) = 1 \quad \text{and}$$

$$L(n+1) = \min \{ j > L(n) \,;\, X_j > X_{L(n)} \} , \quad n \in \mathbb{N} ,$$

are called record times and

$$X_{L(n)} , \; n \in \mathbb{N} ,$$

are called (upper) record values.

Record values have been extensively investigated and only some of the important works in this area are cited here.

Chandler (1952) shows several properties of record values and notes their Markovian structure. Rényi (1962), Tata (1969), Shorrock (1972) and Resnick (1973a) make significant contributions on asymptotics, limit laws, extreme value theory, the structure of record values and characterizations of distributions. A first review on record values is given by Glick (1978) including several applications. Another survey can be found in the book by Galambos (1978) who shows elementary and asymptotic properties of record values and focusses on extreme value theory. Nagaraja (1978) is concerned with conditions for the existence of moments of record values as well as with bounds for expected records which are reviewed in Arnold, Balakrishnan (1989). Pfeifer (1982a), Lin (1987), Witte (1988) and many other authors show characterizations of distributions by means of moments, identical distributions and independence of record statistics.

Concerning the history of record values, applications and statistical inference based on records, we refer in particular to Arnold, Balakrishnan, Nagaraja (1992, Ch.9), to the survey articles of Nevzorov (1987) and Nagaraja (1988) and to Houchens (1984).

1.4.2. Based on an iid sequence of r.v.'s $(X_i)_{i\in\mathbb{N}}$ with an absolutely continuous distribution function F and density function f, we obtain the joint density of the first r record values $X_{L(1)},\dots,X_{L(r)}$

$$f^{X_{L(1)},\dots,X_{L(r)}}(x_1,\dots,x_r) = \left(\prod_{i=1}^{r-1} \frac{f(x_i)}{1 - F(x_i)} \right) f(x_r) ,$$

the marginal densities

$$f^{X_{L(r)}}(x) = \frac{1}{(r-1)!} \left(\log \frac{1}{1 - F(x)} \right)^{r-1} f(x) ,$$

and the marginal distribution functions

$$F^{X_{L(r)}}(x) = 1 - (1 - F(x)) \sum_{j=0}^{r-1} \frac{1}{j!} \left(\log \frac{1}{1 - F(x)} \right)^j .$$

Moreover, the record values form a Markov chain with transition probabilities

$$P(X_{L(r)} > t \mid X_{L(r-1)} = s) = \frac{1 - F(t)}{1 - F(s)} , \quad r \geq 2$$

(cf. Chandler 1952, Shorrock 1972, Grudzień, Szynal 1983, Nagaraja 1988).

Thus we obtain stationary transition probabilities in contrast to order statistics when the above expression is raised to the power $n-r+1$.

Record values are closely connected with the occurrence times of a non–homogeneous Poisson process often used in shock models.
In many practical situations it is not the counting process itself, i.e. the numbers of occurrences up to time t, but the sequence of occurrence times that is important.

The following theorem is stated in Gupta, Kirmani (1988).

Theorem 1.4.3. Let $(N(t))_{t\geq 0}$ be a non–homogeneous Poisson process with a continuous mean value function M satisfying $M(t) \longrightarrow \infty$, $t \longrightarrow \infty$, and $M(0) = 0$.

Let $T_0 = 0$ and $T_i = \min \{ t \geq 0 ; N(t) \geq i \}$, $i \in \mathbb{N}$, be the occurrence times of $(N(t))_{t\geq 0}$ and let F be a continuous distribution function with $F(0) = 0$.

Then the sequence of record values and the sequence of occurrence times are identically distributed:

$$(X_{L(n)})_{n\in\mathbb{N}} \sim (T_n)_{n\in\mathbb{N}}$$

iff

$$F(t) = 1 - \exp\{-M(t)\} , \quad t \geq 0 .$$

Hence, the sequence of record values can be viewed as the sequence of occurrence times of some non–homogeneous Poisson process and vice versa, if the distribution function upon which the records are based and the mean value function of the process are connected as shown above.
Through this, we have another possibility to fit a record model. Consider a situation in

which the waiting times between successive shocks (and not the values of the shocks) are of special interest. That is, shocks appear at the occurrence times of some non–homogeneous Poisson process. Applying the above theorem, the waiting times between shocks can be modelled by record differences. The term 'shock model' is used to describe both situations. In other words, we may be interested in the values of successive peak voltages or in the occurrence times of these peaks.

Gupta, Kirmani (1988) provide a survey of the application of non–homogeneous Poisson processes in the modelling of repairable systems (cf. Ascher, Feingold 1984) as well as in the context of minimal repair. Moreover, they point out and discuss the connection between record values and the relevation transform.

By analogy with the transmission rules for aging properties in the case of order statistics, it is shown in Gupta, Kirmani (1988) that the IFR property of the r–th record is ensured by the IFR property of the underlying distribution. Kochar (1990) generalizes this result in the sense of Takahasi (1988) using the same argument. Thus, the IFR property of some record is transmitted to the following one and the DFR property of some record is transmitted to the previous one. For further details we refer to Chapter V.
Enlarged and generalized models of record values are discussed in the following sections.

1.5. k – Records

We now consider situations in which the record values themselves are viewed as 'outliers' and hence second or third largest values are of special interest. Insurance claims in some non–life insurance can be used as an example.
Observing successive k–th largest values in a sequence, Dziubdziela, Kopociński (1976) propose the following model of k–th record values.

Again, let $(X_i)_{i\in\mathbb{N}}$ be a sequence of iid r.v.'s with a continuous distribution function F. Noticing that, for some $k \in \mathbb{N}$, the sequence

$$(X_{j-k+1,j})_{j=k,k+1,\ldots}$$

of k–th largest order statistics is non–decreasing, the intuitive way to introduce k–th records is the following.

Eliminate all repetitions in this sequence to obtain a strictly increasing subsequence, the elements of which are called k–th record values.
The formal approach is to define record times first.

Definition 1.5.1. Let $(X_i)_{i\in\mathbb{N}}$ be an iid sequence of r.v.'s with a continuous distribution function F and let k be a positive integer.
The r.v.'s $L^{(k)}(n)$ given by

$$L^{(k)}(1) = 1\,,$$
$$L^{(k)}(n+1) = \min\{\, j \in \mathbb{N}\,;\, X_{j,j+k-1} > X_{L^{(k)}(n),L^{(k)}(n)+k-1} \}\,,\quad n \in \mathbb{N}\,,$$

are called k–th record times and the quantities

$$X_{L^{(k)}(n),L^{(k)}(n)+k-1} \quad \text{which we denote by } X_{L^{(k)}(n)}\,,\quad n \in \mathbb{N}\,,$$

are termed k–th record values or k–records.

Obviously, we obtain ordinary record values (see previous section) in the case $k = 1$.

Moreover, Nagaraja (1988) points out that k–records with an underlying distribution function F can be viewed as ordinary record values ($k = 1$) based on the distribution function G (minimum distribution) with

$$G(x) = 1 - (1 - F(x))^k\,.$$

In the above definition, record values are introduced via order statistics with random parameters. The record times fix the first parameter respectively, and the parameter k appears in the sample size.

Another possibility for the definition of k–records is shown in Deheuvels (1984). In this case, the record times $\tilde{L}^{(k)}(n)$ given by

$$\tilde{L}^{(k)}(1) = k\,,$$
$$\tilde{L}^{(k)}(n+1) = \min\{\, j > \tilde{L}^{(k)}(n)\,;\, X_{j-k+1,j} > X_{\tilde{L}^{(k)}(n)-k+1,\tilde{L}^{(k)}(n)} \}\,,\quad n \in \mathbb{N}\,,$$

represent the sample sizes corresponding to the k–records

$$X_{\tilde{L}^{(k)}(n)-k+1,\tilde{L}^{(k)}(n)}\ ,\quad n \in \mathbb{N}\,.$$

In Dziubdziela, Kopociński (1976) we find the density of k–records and results on limiting distributions. The Markov property is stated in Deheuvels (1983). Similar to Nagaraja (1978) (k = 1), Grudzień, Szynal (1983) are concerned with conditions for the existence of moments as well as with bounds for moments. Deheuvels (1984) points out that order statistics and k–records have a similar Markovian structure and by this he shows results for k–records by analogy with well–known results for order statistics. Moreover, we find characterizations of geometrical and exponential distributions via related properties of order statistics and k–th record values.
For more details we once again refer to the surveys of Galambos (1978), Nevzorov (1987) and Nagaraja (1988).

1.5.2. Based on a sequence $(X_i)_{i\in\mathbb{N}}$ of iid r.v.'s possessing an absolutely continuous distribution function F and density function f, we obtain the joint density of the k–records $X_{L^{(k)}(1)},\ldots,X_{L^{(k)}(r)}$

$$f^{X_{L^{(k)}(1)},\ldots,X_{L^{(k)}(r)}}(x_1,\ldots,x_r) = k^r \left(\prod_{i=1}^{r-1} \frac{f(x_i)}{1-F(x_i)} \right) (1-F(x_r))^{k-1} f(x_r)\,,$$

the marginal densities

$$f^{X_{L^{(k)}(r)}}(x) = \frac{k^r}{(r-1)!} \left(\log \frac{1}{1-F(x)} \right)^{r-1} (1-F(x))^{k-1} f(x)\,,$$

and the marginal distribution functions

$$F^{X_{L^{(k)}(r)}}(x) = 1-(1-F(x))^k \sum_{j=0}^{r-1} \frac{1}{j!} \left(k \log \frac{1}{1-F(x)} \right)^j .$$

Moreover, the k–th record values form a Markov chain with transition probabilities

$$P(X_{L^{(k)}(r)} > t \mid X_{L^{(k)}(r-1)} = s) = \left(\frac{1-F(t)}{1-F(s)} \right)^k,\quad r \geq 2$$

(cf. Dziubdziela, Kopociński 1976, Grudzień, Szynal 1983, Deheuvels 1984, Nagaraja 1988).

An alternative way to introduce k–th record times is shown in Nevzorov (1986a, 1987). Deheuvels' definition of the quantities $\mathcal{L}^{(k)}(n)$ is replaced by

$$\mathcal{L}^{(k)}(1) = k ,$$
$$\mathcal{L}^{(k)}(n+1) = \min \{ j > \mathcal{L}^{(k)}(n) ; X_j > X_{j-k,j-1} \} , \quad n \in \mathbb{N} .$$

Noticing that the inequality $X_j > X_{j-k,j-1}$ can equivalently be replaced by $X_{j-k+1,j} > X_{j-k,j-1}$, we see that in this definition the non–decreasing sequence $(X_{j-k+1,j})_{j=k,k+1,...}$ of k–th largest order statistics is directly used.

As a generalization of k–records, Nevzorov (1986a,b) presents k_n–records.

Based on a sequence $(k_n)_{n\in\mathbb{N}}$ of positive integers, record times are inductively defined by

$$L(n+1) = \min \{ j > L(n) ; j - k_j + 1 \le R_j \le j \} , \quad n \in \mathbb{N} ,$$

where R_j is the sequential rank of X_j in the sample $X_1,...,X_j$; i.e. $X_{R_j,j} = X_j$ for all $j \in \mathbb{N}$.

Deheuvels, Nevzorov (1994) obtain strong laws of large numbers, central limit theorems, functional laws of the iterated logarithm and strong invariance principles for the number N(t) of records in the time interval (0,t) as well as for the record times. Moreover, they consider martingales based on L(n) and moment properties of the record times.

In Section 1.7. we will introduce another model of k_n–records in connection with a line scheme of r.v.'s with possibly different underlying distributions.

1.6. Pfeifer's Record Model

In his dissertation, Pfeifer establishes a model of record values based on a double sequence of non–identically distributed random variables (cf. Pfeifer 1979, 1982a,b). The distribution of the underlying random variables may change after each record event. Here, interventions can be modelled as modifying the situation after the occurrence of a record. Used as a shock model, each shock is allowed to have an influence on the magnitude of the subsequent one.

Ordinary record values or k–th records (cf. Section 1.4., 1.5.) are based on a sequence of iid r.v.'s. The assumption of identical distributions is weakened in Pfeifer's model. In relation to other modifications and approaches, we refer to the survey article by Nevzorov (1987, p 214–220).

DEFINITION 1.6.1. (Pfeifer 1979)

Let $\{X_j^{(n)}\}_{n,j\in\mathbb{N}}$ be a double sequence of independent random variables defined on a probability space $(\Omega,\mathcal{A},P)$ with

$$P^{X_j^{(n)}} = P^{X_1^{(n)}}, \quad n, j \in \mathbb{N}.$$

Then inter record times are inductively defined by

$$\Delta_1 = 1,$$
$$\Delta_{n+1} = \min\{\, j \in \mathbb{N}\, ;\, X_j^{(n+1)} > X_{\Delta_n}^{(n)} \}, \quad n \in \mathbb{N},$$

and record values by $X_{\Delta_n}^{(n)}, \; n \in \mathbb{N}$.

This model is extensively examined in Pfeifer (1979, 1982a,b).
E.g., it is shown that, under suitable conditions, the sequence of jump times of an elementary pure birth process and the sequence of record values are identically distributed. It should be noted that the numbering of records is modified here. In the special case of ordinary record values it coincides with the notation of Nevzorov (1987) (cf. 2.6.).

REMARK 1.6.2. Let F_i ($i = 1,\dots,n$) be the underlying distribution function until the i–th record occurs, respectively. Here, $F_1,\dots,F_n$ are supposed to be absolutely continuous. The density functions are denoted by f_i. In Pfeifer's model the joint density function of the records $X_{\Delta_1}^{(1)}, \dots, X_{\Delta_n}^{(n)}$

is then given by $$f^{X_{\Delta_1}^{(1)},\dots,X_{\Delta_n}^{(n)}}(x_1,\dots,x_n) = \left(\prod_{i=1}^{n-1} \frac{f_i(x_i)}{1 - F_{i+1}(x_i)} \right) f_n(x_n).$$

Choosing $F_1 = \dots = F_n$, this is the joint density of ordinary record values (cf. 1.4.2.).

As an example for the use of this record model, suppose that the $X_j^{(r)}$, $j \geq 1$, correspond to random shocks attacking a component which works without failure unless a shock greater than $X_{\Delta_{r-1}}^{(r-1)}$ occurs, e.g. a peak of voltage. Following, a modified component is used for safety reasons enduring shocks up to the magnitude of the last record $X_{\Delta_r}^{(r)}$. Now suppose the distribution function of the subsequent shocks $X_j^{(r+1)}$, $j \geq 1$, to be given by F_{r+1} (possibly) different from F_r.

Pfeifer's record scheme :

$r = 1$	$X_1^{(1)} \equiv X_{\Delta_1}^{(1)}$ (Δ_1)	$\bullet$	$\bullet$	$\dots$			$\sim F_1$
$r = 2$	$X_1^{(2)}$	$X_2^{(2)}$	$X_3^{(2)}$	$\dots$	$X_{\Delta_2}^{(2)}$ (Δ_2)	$\dots$	$\sim F_2$
$r = 3$	$X_1^{(3)}$	$X_2^{(3)}$	$X_3^{(3)}$	$\dots$	$X_{\Delta_3}^{(3)}$ (Δ_3)	$\dots$	$\sim F_3$
$\vdots$	$\vdots$	$\vdots$	$\vdots$				$\vdots$
$r = n$	$X_1^{(n)}$	$X_2^{(n)}$	$X_3^{(n)}$	$\dots$	$X_{\Delta_n}^{(n)}$ (Δ_n)	$\dots$	$\sim F_n$

REMARK 1.6.3. The records $X_{\Delta_j}^{(j)}$, $j \in \mathbb{N}$, form a Markov chain with transition probabilities

$$P(X_{\Delta_j}^{(j)} \leq t \mid X_{\Delta_{j-1}}^{(j-1)} = s) = \frac{F_j(t) - F_j(s)}{1 - F_j(s)} \quad \text{a.s., } t \geq s \text{ , } F(s) < 1 \text{ , } j \geq 2$$

(cf. Pfeifer 1979, p 23).

In this model, characterizations of the exponential distribution are obtained as well as characterizations of the geometric distribution in the case of discrete distribution functions F_i.
Let $\exp(\lambda)$ denote the exponential distribution with density function

$$f_\lambda(x) = \lambda e^{-\lambda x} \quad , x > 0 \, , \lambda > 0 \, .$$

THEOREM 1.6.4. (Pfeifer 1979, 1982a)
Let $X_1^{(n)}$ be $\exp(\lambda_n)$–distributed, $n \geq 2$, with $\lambda_n > 0$.

Then the family $\{X_{\Delta_n}^{(n)}\}_{n \in \mathbb{N}}$ possesses independent, exponentially distributed increments, and $X_{\Delta_n}^{(n)} - X_{\Delta_{n-1}}^{(n-1)}$ is $\exp(\lambda_n)$–distributed, $n \geq 2$.

Under suitable conditions, this turns out to be a characterizing property of the exponential distribution.

THEOREM 1.6.5. (Pfeifer 1979, 1982a)
Let F_1 be strictly increasing on $x \geq 0$, $F_n^{-1}(0) = 0$ and $F_n^{-1}(1) = \infty$, $n \in \mathbb{N}$.
If the family $\{X_{\Delta_n}^{(n)}\}_{n \in \mathbb{N}}$ possesses independent increments, then there exists a sequence $\{\lambda_n\}_{n \geq 2} \subset \mathbb{R}^{>0}$ such that $X_1^{(n)}$ is $\exp(\lambda_n)$–distributed, $n \geq 2$.

Applying the following lemma (cf. Pfeifer 1982a, Deheuvels 1984), a result concerning the independence of $X_{\Delta_{n-1}}^{(n-1)}$ and $X_{\Delta_n}^{(n)} - X_{\Delta_{n-1}}^{(n-1)}$ is obtained.

LEMMA 1.6.6. (Pfeifer 1982a, Deheuvels 1984)

Let F be a distribution function and X, Y random variables satisfying

$$P(X > t \mid Y = s) = \left(\frac{1 - F(t)}{1 - F(s)}\right)^j \quad P^X\text{– a.s.}$$

for some $j \in \mathbb{N}$, $F(s) < 1$, $t \geq s$.

Then we have :

$X - Y$ and Y are independent iff $\frac{1 - F(t+s)}{1 - F(s)}$ does not depend on s .

PROOF If $X - Y$ and Y are independent, then

$P(X - Y > t) = P(X - Y > t \mid Y = s) = P(X > t + s \mid Y = s) = (\frac{1 - F(t+s)}{1 - F(s)})^j$

obviously does not depend on s.

If, on the other hand, $P(X - Y \leq t \mid Y = s) = P(X - Y \leq t)$ for all t , then

$$P(X - Y \leq t , Y \leq s) = \int_{-\infty}^{s} P(X - Y \leq t \mid Y = z)\, dP^Y(z) = P(X - Y \leq t)\, P(Y \leq s) .$$

E.g., let F be the distribution function of an exponential distribution with parameter λ ;

then we have $$\frac{1 - F(t+s)}{1 - F(s)} = e^{-\lambda t} ;$$

thus, $X - Y$ and Y are stochastically independent.

THEOREM 1.6.7 (Pfeifer 1982a)

Let $n \geq 2$. If F_n is the distribution function of an exponential distribution, then $X_{\Delta_{n-1}}^{(n-1)}$ and $X_{\Delta_n}^{(n)} - X_{\Delta_{n-1}}^{(n-1)}$ are independent.

On the other hand, under the assumptions $F_j^{-1}(0) = 0$, $F_j^{-1}(1) = \infty$, $1 \leq j \leq n$, and F_{n-1} strictly increasing on $x > 0$, the independence implies that F_n is the distribution function of an exponential distribution.

This theorem is a generalization of Tata's result for record values (Tata 1969). For a discussion on the strict monotonicity of F_{n-1} we refer to Pfeifer (1979, p 32 or 1982a, p 131).

Choosing $$F_n(x) = 1-(1-F(x))^{\lambda_n}, \; \lambda_n \in \mathbb{N},$$
in the general record model, F_n corresponds to an $\exp(\lambda\cdot\lambda_n)$–distribution, if F is the distribution function of an $\exp(\lambda)$–distribution.

Moreover, the transformation
$$R(x) = -\log(1-F(x))$$
may be useful in the case of a continuous distribution function F (see Resnick 1973a,b), since $R(X_n)$ is exponentially distributed with parameter λ_n, if X_n is distributed according to F_n; then the records of the family $\{R(X_k^{(n)})\}_{n,k}$ coincide with $R(X_{\Delta_n}^{(n)})$, $n \in \mathbb{N}$, a.s.

Pfeifer (1979, p 81) shows that the distribution of the n–th record $X_{\Delta_n}^{(n)}$, with $\lambda_i = i + \nu$ for all $i \in \mathbb{N}$ and $\nu \in \mathbb{N}_0$ fixed, coincides with the distribution of the n–th order statistic based on $n + \nu$ iid random variables with distribution function F, letting F be the standard exponential distribution function (without loss of generality).

Deheuvels (1984) and Gupta (1984) point out that in the case of
$$F_i(x) = 1-(1-F(x))^{n-i+1}, \; i = 1,\dots,n,$$
the Markovian structures of records and ordinary order statistics coincide.
If F is absolutely continuous, then
$$f^{X_{\Delta_1}^{(1)},\dots,X_{\Delta_n}^{(n)}}(x_1,\dots,x_n) = n! \prod_{i=1}^{n} f(x_i) = f^{Y_{1,n},\dots,Y_{n,n}}(x_1,\dots,x_n)$$
where $Y_{i,n}$ is the i–th order statistic from iid r.v.'s $Y_1,\dots,Y_n$ with distribution function F (see Remark 3.3.1.).

Record values are obviously contained in Pfeifer's model in the distribution theoretical sense. Choosing $F_1 = \dots = F_n \; (= F)$, the joint density (see 1.6.2.) is that of record values with distribution function F (see 1.4.2.).
Pfeifer's records as well as sequential order statistics (cf. Section 1.3.) have a Markovian structure. Comparing the corresponding transition probabilities, we observe just the same

structure. That is, applying the transformation

$$G_r(x) = 1-(1-F_r(x))^{n-r+1},$$

we have identical models with respect to distribution theory.

At this point some questions arise as in the case of sequential order statistics.
E.g., is it possible to generalize more well known properties of records? Do we have transmission results for aging properties in this general set–up?
However, as previously mentioned in Section 1.3., the general setting turns out to be too extensive. Thus, we are looking for a suitable subclass again.

1.6.8. In the following, we restrict ourselves to a particular choice of the distribution functions $F_1,\ldots,F_n$ in each line of the scheme of r.v.'s, namely

$$F_r(t) = 1-(1-F(t))^{\beta_r}, \quad 1 \le r \le n,$$

with some distribution function F and positive real numbers $\beta_1,\ldots,\beta_n$ (cf. 1.3.4. for sequential order statistics).

We will see in the sequel that many well known properties of ordinary record values can be generalized to Pfeifer's records. Thus, we have a large class of models at our disposal to describe and to analyse, e.g., certain shock models.

1.7. k_n– Records from Non–Identical Distributions

In the previous section Pfeifer's records are introduced and in Section 1.5. we deal with k–records. Of course, both models can be combined to obtain an enlarged model which may provide a better fit in practical application.
More generally, we start with a given sequence $(k_n)_{n\in\mathbb{N}}$ of positive integers.

In the line scheme of r.v.'s with distribution function F_r in line r, we fix the k_1–th largest value in line 1, the k_2–th largest value in line 2 which is required to be larger than the previous one, etc.

DEFINITION 1.7.1. Let $\{X_j^{(n)}\}_{n,j\in\mathbb{N}}$ be a double sequence of independent r.v.'s and let $F_1, F_2, \ldots$ be distribution functions with

$$X_j^{(n)} \sim X_1^{(n)} \sim F_n\,,\quad n,j\in\mathbb{N}\,.$$

Moreover, let $(k_n)_{n\in\mathbb{N}}$ be a sequence of positive integers.

Then inter record times are defined by

$$\Delta_1 = 1\,,$$
$$\Delta_{n+1} = \min\{\, j\in\mathbb{N}\,;\, X^{(n+1)}_{j,\,j+k_{n+1}-1} > X^{(n)}_{\Delta_n,\Delta_n+k_n-1}\}\,,\quad n\in\mathbb{N}\,,$$

and record values by

$$X^{(n)}_{\Delta_n,\Delta_n+k_n-1} \quad \text{which we denote by } X^{(n)}_{\Delta_n,k_n}\,.$$

In the case $k_n = 1$ for all $n\in\mathbb{N}$, we obtain Pfeifer's records.

If all distribution functions are identical, this model may be viewed as another generalization of k–records (see Section 1.5.) in addition to the proposal of Nevzorov (1986a).

1.7.2. If we assume that the distribution functions $F_1, F_2, \ldots$ are absolutely continuous with density functions $f_1, f_2, \ldots$ respectively, then it can be shown that the joint density of the first r records $X^{(1)}_{\Delta_1,k_1}, \ldots, X^{(r)}_{\Delta_r,k_r}$ is given by

$$f^{X^{(1)}_{\Delta_1,k_1},\ldots,X^{(r)}_{\Delta_r,k_r}}(x_1,\ldots,x_r) = \left(\prod_{j=1}^{r} k_j\right) \prod_{i=1}^{r} \left(\frac{1-F_i(x_i)}{1-F_i(x_{i-1})}\right)^{k_i-1} \frac{f_i(x_i)}{1-F_i(x_{i-1})}\,,\quad x_0 = -\infty\,.$$

This can be seen via induction on r applying the following lemmas.

LEMMA 1.7.3. Let $X_{1,n}, \ldots, X_{n,n}$ be order statistics based on an absolutely continuous distribution function F. Then we have

$$P(X_{r-1,n} \le s\,,\, X_{r,n} > t) = \binom{n}{r-1}(1-F(t))^{n-r+1}F^{r-1}(s)\,,\quad s<t\,,\quad 1\le r\le n\,,\quad X_{0,n} = -\infty\,.$$

LEMMA 1.7.4. Let $(X_i)_{i\in\mathbb{N}}$ be a sequence of iid r.v.'s with an absolutely continuous distribution function F, let $k \in \mathbb{N}$ be arbitrary, but fixed, and let

$$\tau = \tau_F = \min \{ j \in \mathbb{N} ; X_{j,j+k-1} > s \} , \quad s \in \mathbb{R} .$$

Then for $s < t$ we have

$$P(X_{\tau,\tau+k-1} > t) = \left(\frac{1 - F(t)}{1 - F(s)} \right)^k , \quad t < F^{-1}(1) .$$

1.7.5. Moreover, the k_n–records $X^{(1)}_{\Delta_1,k_1}, X^{(2)}_{\Delta_2,k_2}, \ldots$ form a Markov chain with transition probabilities

$$P(X^{(r)}_{\Delta_r,k_r} > t \mid X^{(r-1)}_{\Delta_{r-1},k_{r-1}} = s) = \left(\frac{1 - F_r(t)}{1 - F_r(s)} \right)^{k_r} , \quad r \geq 2 .$$

In the distribution theoretical sense, the k_n–record model and Pfeifer's model are identical, since the following transformation can be used:

$$G_r(t) = 1 - (1 - F_r(t))^{k_r} .$$

However, the interpretations of the models are obviously different.
Putting $k_r = n - r + 1$, we obtain the Markovian structure of sequential order statistics.

1.7.6. By analogy with 1.6.8., we restrict ourselves to a particular choice of the distribution functions $F_1,\ldots,F_n$ in each line of the scheme of r.v.'s, namely

$$F_r(t) = 1 - (1 - F(t))^{\beta_r} , \quad 1 \leq r \leq n ,$$

with some distribution function F and positive real numbers $\beta_1,\ldots,\beta_n$.

1.8. Ordering via Truncation of Distributions

The model of sequential order statistics (cf. Section 1.3.) is motivated by some (n–r+1)–out–of–n–system with an inherent dependence structure. After the occurrence of the i–th failure in the system at time x, say, the next failure time is modelled as the minimum from a sample of size $n-i$ of iid r.v.'s with a possibly different distribution function F_{i+1} truncated on the left at x. Therefore, in the line scheme

$$(Z_j^{(i)})_{1\le i\le n,\ 1\le j\le n-i+1}$$

of the underlying r.v.'s, line i consists of $n-i+1$ r.v.'s.

Apart from this interpretation, we now consider line schemes with k_i r.v.'s in line i and are interested in the line minima.
To obtain ordered quantities, the distribution functions $F_1,\dots,F_n$ corresponding to the lines $1,\dots,n$ respectively, are appropriately truncated on the left.

We adopt Definition 1.3.1. (sequential order statistics) with a slight modification to define

such ordered r.v.'s $$X_{*,k_1}^{(1)} \le \dots \le X_{*,k_n}^{(n)} .$$

Proceeding in this way, we point out that there are other interpretations of record values.

DEFINITION 1.8.1. Let $k_1,\dots,k_n$ be positive integers and let

$$(Y_j^{(i)})_{1\le i\le n,\ 1\le j\le k_i}$$

be independent r.v.'s with $$(Y_j^{(i)})_{1\le j\le k_i} \sim F_i\,,\quad 1\le i\le n\,,$$

where $$F_1\,,\dots,\,F_n$$

are strictly increasing and continuous distribution functions

with $$F_1^{-1}(1) \le \dots \le F_n^{-1}(1)\,.$$

Then the r.v.'s $$X_{*,k_1}^{(1)}\,,\dots,\,X_{*,k_n}^{(n)}$$

are defined by $$X_{*,k_0}^{(0)} = -\infty\,,$$

$$X_j^{(i)} = F_i^{-1}\Big(\,F_i(Y_j^{(i)})\,(1-F_i(X_{*,k_{i-1}}^{(i-1)})) + F_i(X_{*,k_{i-1}}^{(i-1)}\,)\Big)\,,\quad 1\le i\le n\,,$$

$$X_{*,k_i}^{(i)} = \min\,\{\,X_j^{(i)},\ 1\le j\le k_i\,\}\,.$$

That is, the r.v.'s $X_{*,k_i}^{(i)}$, $1 \leq i \leq n$, are line minima in the scheme

$$(X_j^{(i)})_{1 \leq i \leq n,\ 1 \leq j \leq k_i}$$

of r.v.'s defined above.

By analogy with the results for sequential order statistics (cf. 1.3.3.) we find :

1.8.2. If the distribution functions $F_1,\dots,F_n$ are absolutely continuous with corresponding density functions $f_1,\dots,f_n$, then the joint density of the first r line minima $X_{*,k_1}^{(1)}$, ... , $X_{*,k_r}^{(r)}$ is given by

$$f^{X_{*,k_1}^{(1)},\dots,X_{*,k_r}^{(r)}}(x_1,\dots,x_r) = \left(\prod_{j=1}^{r} k_j\right) \prod_{i=1}^{r} \left(\frac{1 - F_i(x_i)}{1 - F_i(x_{i-1})}\right)^{k_i - 1} \frac{f_i(x_i)}{1 - F_i(x_{i-1})}, \quad x_0 = -\infty .$$

The quantities form a Markov chain with transition probabilities

$$P(\, X_{*,k_r}^{(r)} > t \mid X_{*,k_{r-1}}^{(r-1)} = s\,) = \left(\frac{1 - F_r(t)}{1 - F_r(s)}\right)^{k_r}, \quad 2 \leq r \leq n .$$

That is, the above simple construction applied to obtain ordered r.v.'s via truncation of distributions leads to alternative interpretations of the record models which are shown in the previous sections.

REMARKS 1.8.3.

i) The Markovian structure of $X_{*,k_1}^{(1)}$, ... , $X_{*,k_n}^{(n)}$ coincides with that of k_n–records from non–identical distributions (Section 1.7.).

Hence, in the distribution theoretical sense, these models are identical and the records may be interpreted as certain line minima.

ii) Putting $k_i = n - i + 1, \ 1 \le i \le n,$

we get the structure of sequential order statistics ($\prod_{j=1}^{r} k_j = \frac{n!}{(n-r)!}$).

iii) Choosing $k_i = 1$ for all i

yields Pfeifer's record model (cf. Section 1.6.).

That is, record values from non–identically distributed r.v.'s can simply be viewed as suitably transformed r.v.'s $Y_1^{(i)}$, $1 \le i \le n$, which are distributed according to F_i, respectively:

$$X_{*,1}^{(1)} = Y_1^{(1)},$$

$$X_{*,1}^{(i)} = F_i^{-1}\left(F_i(Y_1^{(i)})\,(1 - F_i(X_{*,1}^{(i-1)})) + F_i(X_{*,1}^{(i-1)}) \right), \quad 2 \le i \le n .$$

iv) In the case of identical distribution functions $F_1 = \ldots = F_n$, we obtain alternative interpretations of ordinary record values ($k_i = 1$, $1 \le i \le n$, cf. Section 1.4.) and of k–records ($k_i = k \in \mathbb{N}$, $1 \le i \le n$, cf. Section 1.5.).

1.8.4. By analogy with 1.3.4., we restrict ourselves to a particular choice of the distribution functions $F_1,\ldots,F_n$, namely

$$F_r(t) = 1 - (1 - F(t))^{\alpha_r}, \quad 1 \le r \le n,$$

with some distribution function F and positive real numbers $\alpha_1,\ldots,\alpha_n$.

1.9. Censoring Schemes

Censored data are considered, e.g., in reliability theory, survival analysis and clinical trials. A survey of this topic as well as a comprehensive bibliography are presented by McCool (1982).

Several models of censoring can be described in terms of order statistics. Their use for statistical inference in models of censored data is shown in Mann, Schafer, Singpurwalla (1974, Ch.5), David (1981, Ch.6), Lawless (1982, Ch.1), Nelson (1982, Ch.7), Balakrishnan, Cohen (1991, Ch.10) and Arnold, Balakrishnan, Nagaraja (1992, Ch.7).

In this section we consider progressive type II censoring with two stages as proposed by Lawless (1982, p 33/4).
The following situation is modelled by means of a random sample of ν iid lifetimes $X_1,\dots,X_\nu$ with density function f and distribution function F.

At the time of the r_1–th failure in a sample of ν items, n_1 of the remaining $\nu - r_1$ unfailed items are randomly selected and removed from the experiment. Then $\nu - r_1 - n_1$ items are still present. The experiment terminates when further r_2 items have failed.

Thus we have r_1 observations $x_1 \le \dots \le x_{r_1}$ in the first stage of the experiment and r_2 failure times $x_{r_1+1} \le \dots \le x_{r_1+r_2}$ in the second stage.

Putting $n = r_1 + r_2$, Lawless (1982) obtains the corresponding likelihood function as

$$\frac{\nu! \quad (\nu-r_1-n_1)!}{(\nu-r_1)! \; (\nu-n_1-n)!} \left(\prod_{i=1}^{n} f(x_i) \right) (1-F(x_{r_1}))^{n_1} (1-F(x_n))^{\nu-n_1-n} , \quad x_1 \le \dots \le x_n .$$

The above joint density function coincides with the joint density of n sequential OS's choosing

$$\alpha_i = \begin{cases} \dfrac{\nu-i+1}{n-i+1} , & 1 \le i \le r_1 \\[2ex] \dfrac{\nu-n_1-i+1}{n-i+1} , & r_1 < i \le n \end{cases}$$

(cf. 1.3.3., 1.3.4.) as well as with the joint density of n records in Pfeifer's model choosing

$$\beta_i = \begin{cases} \nu-i+1 , & 1 \le i \le r_1 \\ \nu-n_1-i+1 , & r_1 < i \le n \end{cases}$$

(cf. 1.6.2., 1.6.8.).

Therefore, in the distribution theoretical sense, the outcomes in the situation of progressive type II censoring with two stages can be viewed as certain sequential OS's or as record values from non–identical distributions.

2. The Definition of Generalized Order Statistics and Examples

In Section 1 several models of ordered r.v.'s are introduced, including order statistics and record values. In the distribution theoretical sense, all models, or appropriately restricted versions of these models, are contained in the model of generalized order statistics which we now establish.

First, we introduce uniform generalized order statistics. Using the quantile transformation, we then define generalized order statistics based on an arbitrary distribution function F.

The starting point is a joint density function on a cone of the n–dimensional Euclidean space. Assuming an absolutely continuous distribution function F and choosing the parameters in the joint density of generalized order statistics appropriately, we obtain the densities of ordered r.v.'s shown in the Sections 1.1. – 1.9.

DEFINITION 2.1.

Let $n \in \mathbb{N}$, $k \geq 1$, $m_1, \ldots, m_{n-1} \in \mathbb{R}$, $M_r = \sum_{j=r}^{n-1} m_j$, $1 \leq r \leq n-1$, be parameters such that

$$\gamma_r = k + n - r + M_r \geq 1 \quad \text{for all} \quad r \in \{1,\ldots,n-1\}$$

and let $\tilde{m} = (m_1,\ldots,m_{n-1})$, if $n \geq 2$ ($\tilde{m} \in \mathbb{R}$ arbitrary, if $n = 1$).

If the random variables $\quad U(r,n,\tilde{m},k)$, $r = 1,\ldots,n$,

possess a joint density function of the form

$$f^{U(1,n,\tilde{m},k),\ldots,U(n,n,\tilde{m},k)}(u_1,\ldots,u_n) = k\left(\prod_{j=1}^{n-1} \gamma_j\right)\left(\prod_{i=1}^{n-1} (1-u_i)^{m_i}\right)(1-u_n)^{k-1}$$

on the cone $\quad 0 \leq u_1 \leq \ldots \leq u_n < 1$

of $\mathbb{R}^n$, then they are called

uniform generalized order statistics .

In the particular case $m_1 = \ldots = m_{n-1} = m$, say, the random variables are denoted by

$$U(r,n,m,k),\ r = 1,\ldots,n .$$

REMARK 2.2. In the definition of uniform generalized order statistics it is required that

$$k \geq 1 \quad \text{and} \quad \gamma_r \geq 1 \,, \quad 1 \leq r \leq n-1 \,.$$

In order to develop the distribution theory of uniform generalized order statistics (e.g. marginal density functions or marginal distribution functions), it is sufficient to assume

$$k > 0 \quad \text{and} \quad \gamma_r > 0 \,, \quad 1 \leq r \leq n-1 \,.$$

However, in connection with finite upper bounds for the marginal densities (see 3.2.12.) and with conditions for the existence of moments (see II. 1.1.3.,1.1.4.), the stronger assumption turns out to be useful.

An important special case in the concept of g OS's is choosing $\tilde{m}$ according to

$$m_1 = \dots = m_{n-1} = m \in \mathbb{R} \,.$$

Here, the main regularity conditions imposed on k, n and $r \leq n$ are

$$k \geq 1 \quad \text{and} \quad k + (n-1)(m+1) \geq 1$$

which imply $\quad k + (n-r)(m+1) \geq 1 \quad \text{for } r \in \{1,\dots,n\} \,.$

Generalized order statistics based on some distribution function F are now defined by means of the quantile transformation

$$X(r,n,\tilde{m},k) = F^{-1}(\, U(r,n,\tilde{m},k)\,) \,, \quad r = 1,\dots,n \,.$$

DEFINITION 2.3.

Let the situation of Definition 2.1. be given and let F be a distribution function.

The random variables

$$X(r,n,\tilde{m},k) = F^{-1}(\, U(r,n,\tilde{m},k)\,) \,, \quad r = 1,\dots,n \,,$$

are called **generalized order statistics**

(based on the distribution function F).

In the sequel we write ' g OS's ' for brevity.

In the case $m_1 = \dots = m_{n-1} = m$, say, they are denoted by

$$X(r,n,m,k) \,, \; r = 1,\dots,n \,.$$

REMARK 2.4. Let the distribution function F in Definition 2.3. be absolutely continuous with density function f.

Noticing that

$$\begin{aligned} & P(\, X(1,n,\tilde{m},k) \le x_1 \,, \dots , X(n,n,\tilde{m},k) \le x_n \,) \\ &= P(\, F^{-1}(U(1,n,\tilde{m},k)) \le x_1 \,, \dots , F^{-1}(U(n,n,\tilde{m},k)) \le x_n \,) \\ &= P(\, U(1,n,\tilde{m},k)) \le F(x_1) \,, \dots , U(n,n,\tilde{m},k)) \le F(x_n) \,) \,, \end{aligned}$$

the joint density function of the generalized order statistics

$$X(1,n,\tilde{m},k) \,, \dots , X(n,n,\tilde{m},k)$$

is given by

$$\begin{aligned} & f^{X(1,n,\tilde{m},k),\dots,X(n,n,\tilde{m},k)}(x_1,\dots,x_n) \\ &= k \,(\prod_{j=1}^{n-1} \gamma_j) \left(\prod_{i=1}^{n-1} (1-F(x_i))^{m_i} f(x_i) \right) (1-F(x_n))^{k-1} f(x_n) \end{aligned}$$

on the cone

$$F^{-1}(0) < x_1 \le \dots \le x_n < F^{-1}(1) \,.$$

Order statistics from a sample of iid r.v.'s as well as record values based on a sequence of iid r.v.'s are obviously included in the distribution theoretical sense.

Not only these models, but all the structures introduced in Section 1 are contained in the model of g OS's. That is, choosing the parameters in 2.3. appropriately, we obtain the corresponding joint density functions shown in Section 1.

EXAMPLES 2.5.

i) In the case $m_1 = \dots = m_{n-1} = 0$ and $k = 1$ (i.e. $\gamma_r = n - r + 1 \,,\ 1 \le r \le n-1$), the model reduces to the joint density of o OS's from n iid r.v.'s $X_1,\dots,X_n$ with distribution function F :

$$f^{X(1,n,0,1),\dots,X(n,n,0,1)}(x_1,\dots,x_n) = n! \prod_{i=1}^{n} f(x_i)$$

(e.g. David 1981, p 10; see Section 1.1.).

ii) Choosing $m_1 = \dots = m_{n-1} = 0$ and $k = \alpha - n + 1$ with $n - 1 < \alpha \in \mathbb{R}$ (i.e. $\gamma_r = \alpha - r + 1 \,,\ 1 \le r \le n-1$), we describe OS's with non–integral sample size :

$$f^{X(1,n,0,\alpha-n+1),\dots,X(n,n,0,\alpha-n+1)}(x_1,\dots,x_n)$$

$$= \Big(\prod_{j=1}^{n} (\alpha-j+1)\Big)\,(1-F(x_n))^{\alpha-n} \prod_{i=1}^{n} f(x_i)$$

(see Section 1.2.).

iii) Given positive real numbers $\alpha_1,\dots,\alpha_n$, we put

$$m_i = (n-i+1)\,\alpha_i - (n-i)\,\alpha_{i+1} - 1\ ,\quad i = 1,\dots,n-1 \ \text{ and } \ k = \alpha_n$$

(i.e. $\gamma_r = (n-r+1)\,\alpha_r\ ,\ 1 \le r \le n-1$) to obtain the joint density

$$f^{X(1,n,\tilde{m},\alpha_n),\dots,X(n,n,\tilde{m},\alpha_n)}(x_1,\dots,x_n)$$

$$= n!\,\Big(\prod_{j=1}^{n} \alpha_j\Big)\left(\prod_{i=1}^{n-1} (1-F(x_i))^{m_i} f(x_i)\right)(1-F(x_n))^{\alpha_n-1} f(x_n)$$

of sequential order statistics based on

$$F_r(t) = 1-(1-F(t))^{\alpha_r}\ ,\quad 1 \le r \le n\ ,$$

where F is an arbitrary, absolutely continuous distribution function (see Section 1.3., in particular 1.3.4.).

iv) In the case $m_1 = \dots = m_{n-1} = -1$ and $k \in \mathbb{N}$

(i.e. $\gamma_r = k\ ,\ 1 \le r \le n-1$), we obtain the joint density of the first n k–th record values based on the sequence $(X_i)_{i\in\mathbb{N}}$ of iid r.v.'s with distribution function F :

$$f^{X(1,n,-1,k),\dots,X(n,n,-1,k)}(x_1,\dots,x_n)$$

$$= k^n \left(\prod_{i=1}^{n-1} \frac{f(x_i)}{1-F(x_i)}\right)(1-F(x_n))^{k-1} f(x_n)$$

(e.g. Chandler 1952, Nagaraja 1988; see Sections 1.4. for $k = 1$ and 1.5. for $k \in \mathbb{N}$).

The term 'k–th record values' is used in the literature. For this reason the parameter k is chosen in 2.1. instead of $m_n + 1$.

v) Given positive real numbers $\beta_1,\dots,\beta_n$, we choose

$$m_i = \beta_i - \beta_{i+1} - 1\,,\ i = 1,\dots,n-1 \ \text{ and } \ k = \beta_n$$

(i.e. $\gamma_r = \beta_r$, $1 \le r \le n-1$) to obtain the joint density

$$f^{X(1,n,\tilde{m},\beta_n),\dots,X(n,n,\tilde{m},\beta_n)}(x_1,\dots,x_n)$$

$$= \Big(\prod_{j=1}^{n} \beta_j\Big) \Big(\prod_{i=1}^{n-1} (1-F(x_i))^{m_i} f(x_i)\Big) (1-F(x_n))^{\beta_n - 1} f(x_n)$$

of Pfeifer's record values from non–identically distributed r.v.'s based on

$$F_r(t) = 1-(1-F(t))^{\beta_r}\,,\ 1 \le r \le n\,,$$

where F is an arbitrary, absolutely continuous distribution function (see Section 1.6., in particular 1.6.8.).

vi) By analogy with v), we choose

$$m_i = \beta_i \cdot k_i - \beta_{i+1} \cdot k_{i+1} - 1\,,\ i = 1,\dots,n-1 \ \text{ and } \ k = \beta_n \cdot k_n\,,$$

$k_1,\dots,k_n \in \mathbb{N}$ (i.e. $\gamma_r = \beta_r \cdot k_r$, $1 \le r \le n-1$) to obtain the joint density of k_n–records from non–identically distributed r.v.'s based on

$$F_r(t) = 1-(1-F(t))^{\beta_r}\,,\ 1 \le r \le n$$

(see Section 1.7.).

In particular, choosing identical distributions $F_r(t) = F(t)$, $1 \le r \le n$ (i.e. $\beta_r = 1$, $1 \le r \le n$, $\gamma_r = k_r$, $1 \le r \le n-1$), we derive the joint density function

$$f^{X(1,n,\tilde{m},k_n),\dots,X(n,n,\tilde{m},k_n)}(x_1,\dots,x_n)$$

$$= \Big(\prod_{j=1}^{n} k_j\Big) \Big(\prod_{i=1}^{n-1} (1-F(x_i))^{k_i - k_{i+1} - 1} f(x_i)\Big) (1-F(x_n))^{k_n - 1} f(x_n)$$

of ordered r.v.'s which can be viewed as a generalization of k–records. The corresponding joint densities coincide in the case $k_1 = \dots = k_n$ ($= k$) (see iv)).

vii) For different interpretations of the record values in iv), v) and vi) we refer to Section 1.8.

viii) Choosing $m_1 = \ldots = m_{r_1-1} = m_{r_1+1} = \ldots = m_{n-1} = 0$,

$$m_{r_1} = n_1 \text{ and } k = \nu - n_1 - n + 1$$

(i.e. $\gamma_r = \nu - r + 1$, $1 \le r \le r_1$, $\gamma_r = \nu - n_1 - r + 1$, $r_1 < r \le n-1$), we obtain the joint density arising in progressive type II censoring with two stages:

$$f^{X(1,n,\tilde{m},k),\ldots,X(n,n,\tilde{m},k)}(x_1,\ldots,x_n)$$

$$= \frac{\nu! \quad (\nu-r_1-n_1)!}{(\nu-r_1)! \; (\nu-n_1-n)!} \left(\prod_{i=1}^{n} f(x_i) \right) (1-F(x_{r_1}))^{n_1} (1-F(x_n))^{\nu-n_1-n}$$

(see Section 1.9.).

ix) Given positive real numbers $\upsilon_0, \upsilon_1, \ldots, \upsilon_{n-1}$, we choose

$$m_i = \frac{n-i+1}{\upsilon_{i-1}} - \frac{n-i}{\upsilon_i} - 1, \quad 1 \le i \le n-1 \text{ and } k = \frac{1}{\upsilon_{n-1}}$$

(i.e. $\gamma_r = \frac{n-r+1}{\upsilon_{r-1}}$, $1 \le r \le n-1$).

Moreover, let F be the distribution function of the standard exponential distribution $F(x) = 1 - e^{-x}$, $x > 0$.

Then the joint density of the corresponding g OS's coincides with the density of Weinman's multivariate exponential distribution which is an extension of Freund's bivariate exponential distribution (see Johnson, Kotz 1972, Ch.41). Suppose that a system consists of n components of the same kind with times to failure $Y_1,\ldots,Y_n$ which are exponentially distributed with parameter $1/\upsilon_0$:

$$f^{Y_1}(y) = \frac{1}{\upsilon_0} e^{-y/\upsilon_0}, \quad y > 0, \quad \upsilon_0 > 0.$$

If r of the components have failed, the conditional joint distribution of the lifetimes of the remaining components is supposed to be that of independent r.v.'s, each having an exponential distribution with parameter $1/\upsilon_r$, $\upsilon_r > 0$.

Weinman showed that the joint density of the ordered quantities $Y'_1 \le \ldots \le Y'_n$ is given by

$$f^{Y'_1,\ldots Y'_n}(y_1,\ldots,y_n) = n! \prod_{i=0}^{n-1} \frac{1}{\upsilon_i} \exp\{-\frac{n-i}{\upsilon_i}(y_{i+1} - y_i)\}$$

where $y_0 = 0$ and $y_1 \leq \dots \leq y_n$.

Choosing the parameters $m_1,\dots,m_{n-1}$ and k as shown above, we obtain

$$f^{Y_1',\dots Y_n'} = f^{X(1,n,\tilde{m},k),\dots,X(n,n,\tilde{m},k)} .$$

Therefore, distributional properties of $Y_1', \dots, Y_n'$ can be described by certain g OS's.

Remark 2.6. When comparing results from the literature, the different numbering of record values has to be taken into account. E.g., Resnick (1973a,b), Nagaraja (1978), Pfeifer (1979, 1982a), Grudzień, Szynal (1983), Lin (1988a,b), Too, Lin (1989) and Gajek, Gather (1991) start counting with 0 ($L^{(k)}(0) = 1$). Whereas in this monograph, the natural numbers are chosen as the set of indices as found in Dziubdziela, Kopociński (1976), Deheuvels (1984), Nevzorov (1987) and Nagaraja (1988).

As previously pointed out in Section 1., the presented models of ordered r.v.'s can be interpreted in different ways. In the above example it is shown that all these models, or appropriately restricted versions (cf. 1.3.4., 1.6.8.) respectively, can be described by means of g OS's in the distribution theoretical sense.
Therefore, we now have an interpretation of order statistics with non–integral sample size.

They can be viewed as sequential order statistics based on

$$F_r(t) = 1 - (1 - F(t))^{\alpha_r}, \quad 1 \leq r \leq n,$$

choosing
$$\alpha_r = \frac{\alpha - n}{n - r + 1} + 1$$

(i.e. $m_i = (n-i+1)\,\alpha_i - (n-i)\,\alpha_{i+1} - 1 = 0$, $i = 1,\dots,n-1$ and $k = \alpha_n = \alpha - n + 1$; cf. 2.4.iii)),

or as Pfeifer's record values based on

$$F_r(t) = 1 - (1 - F(t))^{\beta_r}, \quad 1 \leq r \leq n,$$

putting
$$\beta_r = \alpha - r + 1$$

(i.e. $m_i = \beta_i - \beta_{i+1} - 1 = 0$, $i = 1,\dots,n-1$ and $k = \beta_n = \alpha - n + 1$; cf. 2.4.v)).

REMARK 2.7. On the other hand, g OS's based on F can always be interpreted as sequential order statistics ($F_r(t) = 1-(1-F(t))^{\gamma_r/(n-r+1)}$), Pfeifer's records ($F_r(t) = 1-(1-F(t))^{\gamma_r}$), k_n–records from non–identical distributions ($F_r(t) = 1-(1-F(t))^{\gamma_r/k_r}$) or as some ordered r.v.'s based on truncated distributions ($F_r(t) = 1-(1-F(t))^{\gamma_r/k_r}$; cf. Section 1.8.).

REMARK 2.8. For any joint density function shown in 2.4., g OS's may be constructed via Pfeifer's model of record values (Pfeifer 1979, 1982a,b) based on a double sequence of random variables with

$$F_i(x) = 1-(1-F(x))^{k+n-i+M_i}$$

$m_j \in \mathbb{R}$, $k \geq 1$ and $k+n-i+M_i \geq 1$ for $i = 1,\dots,n$:

$$f^{X_{\Delta_1}^{(1)},\dots,X_{\Delta_n}^{(n)}}(x_1,\dots,x_n) = k\left(\prod_{j=1}^{n-1}\gamma_j\right)\left(\prod_{i=1}^{n-1}(1-F(x_i))^{m_i} f(x_i)\right)(1-F(x_n))^{k-1} f(x_n)\,,$$

$$F^{-1}(0) < x_1 \leq \dots \leq x_n < F^{-1}(1)\,.$$

Thus, Pfeifer's results can be transferred to g OS's; e.g.:

The g OS's $X(r,n,\tilde{m},k)$, $r = 2,\dots,n$, form a Markov chain with transition probabilities ($F(s) < 1$):

$$P(\,X(r,n,\tilde{m},k) > t \mid X(r-1,n,\tilde{m},k) = s\,) = \frac{1-F_r(t)}{1-F_r(s)} = \left(\frac{1-F(t)}{1-F(s)}\right)^{k+n-r+M_r}, \; t \geq s$$

(see Section 1.6.).

In its general version, Pfeifer's model (as well as the model of sequential order statistics) is not suitable for our purposes. It does not provide a common approach to distributional properties and to results on moments of g OS's (see Ch. II, III, IV; cf. Sections 1.3. and 1.6.). Pfeifer's model is too extensive to explain the analogies in the behaviour of order statistics and record values (see Introduction) as well as to generalize known results and to integrate them within a general framework. Therefore, we restrict ourselves to a particular choice of the underlying distributions (cf. 2.5.v), 2.8. and 1.6.8.). However, the Markovian structure of Pfeifer's records does provide a helpful tool leading to a unified approach to a variety of results on independence or identical distributions of certain statistics. Here, these facts are only referred to marginally and asymptotic investigations excluded. For a more detailed discussion we refer to Pfeifer (1979).

REMARK 2.9.
If g OS's are interpreted as record values from non-identically distributed random variables in Pfeifer's model, then successive underlying distributions (i.e. underlying distributions before and after any record event) are always stochastically ordered.
Depending on the choice of $m_1,\ldots,m_{n-1}$, we have transitions to stochastically larger or smaller distributions.

Let $1 \le i \le n-1$. Then we find (cf. Remark 2.8.) :

$$F_i(x) \left\{ \begin{matrix} < \\ = \\ > \end{matrix} \right\} F_{i+1}(x) \quad \text{for all } x \in (F^{-1}(0), F^{-1}(1)) \Leftrightarrow m_i \left\{ \begin{matrix} < \\ = \\ > \end{matrix} \right\} -1 .$$

Examples ($m_1 = \ldots = m_{n-1} = m \Rightarrow \gamma_i = k + n - i + M_i = k + (n-i)(m+1)$, $1 \le i \le n-1$):

i) $m_1 = \ldots = m_{n-1} = -2$, $k = n$

i	1	2	...	$n-1$	n
$k + (n-i)(m+1)$	1	2	...	$n-1$	n

with $F_i(x) < F_{i+1}(x)$ for all $x \in (F^{-1}(0), F^{-1}(1))$, $1 \le i \le n-1$.

ii) $m_1 = \ldots = m_{n-1} = 1$, $k = 1$

i	1	2	...	$n-1$	n
$k + (n-i)(m+1)$	$2(n-1)+1$	$2(n-2)+1$	...	3	1

with $F_i(x) > F_{i+1}(x)$ for all $x \in (F^{-1}(0), F^{-1}(1))$, $1 \le i \le n-1$.

Choosing F (cf. Remark 2.8.) as some Pareto or Weibull distribution function, e.g., then the distribution functions F_i , $1 \le i \le n$, belong to the same family of distributions, respectively.

If F is a Pareto distribution function with parameters α and p:

$$F(x) = 1 - (\tfrac{\alpha}{x})^p , \; x > \alpha > 0 , \; p > 0 ,$$

then the F_i, $1 \le i \le n$, are also Pareto distribution functions and F_i is parametrized by α and $\gamma_i \cdot p$.

If F is a Weibull distribution function with parameters λ and p:

$$F(x) = 1 - \exp(-\lambda\, x^p) , \; x > 0 , \; \lambda, p > 0 ,$$

then the F_i, $1 \le i \le n$, are also Weibull distribution functions and F_i is parametrized by $\gamma_i \cdot \lambda$ and p.

The following diagrams show density functions and distribution functions of Weibull distributions with $\lambda = 1$, $p = 2$, $\gamma_i = i$, $1 \le i \le 5$ ($m_1 = \ldots = m_{n-1} = -2$, $k = n = 5$):

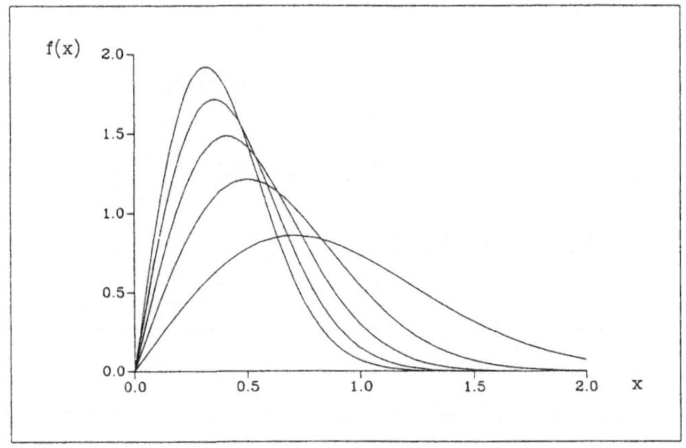

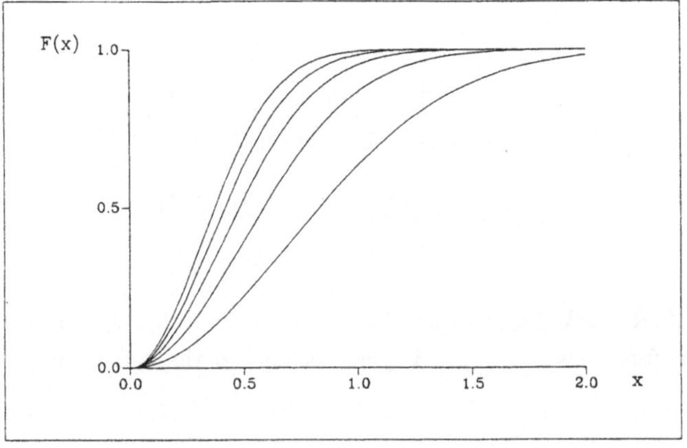

The following table provides a survey on particular models of ordered r.v.'s and their parametrization within the model of g OS's.

Based on an absolutely continuous distribution function F with density function f, the joint density function of the g OS's $X(1,n,\tilde{m},k),\ldots,X(n,n,\tilde{m},k)$ is given by

$$f^{X(1,n,\tilde{m},k),\ldots,X(n,n,\tilde{m},k)}(x_1,\ldots,x_n)$$

$$= k\left(\prod_{j=1}^{n-1}\gamma_j\right)\left(\prod_{i=1}^{n-1}(1-F(x_i))^{m_i}f(x_i)\right)(1-F(x_n))^{k-1}f(x_n),\quad x_1\le\ldots\le x_n,$$

(see 2.4.) which leads to the transition probabilities

$$P(\,X(r,n,\tilde{m},k) > t \mid X(r-1,n,\tilde{m},k) = s\,) = \left(\frac{1-F(t)}{1-F(s)}\right)^{\gamma_r},\quad t\ge s$$

(see 2.8.).

Sec.	m_r, $1\le r\le n-1$	k	γ_r, $1\le r\le n-1$	models
1.1.	0	1	$n-r+1$	order statistics
1.2.	0	$\alpha-n+1$	$\alpha-r+1$	order statistics with non–integral sample size
1.3.	$(n-r+1)\,\alpha_r$ $-(n-r)\,\alpha_{r+1}-1$	α_n	$(n-r+1)\,\alpha_r$	sequential order statistics
1.4.	-1	1	1	record values, occurrence times of non–homogeneous Poisson process, minimal repair, relevation transform
1.5.	-1	$k\in\mathbb{N}$	k	k–th record values
1.6.	$\beta_r-\beta_{r+1}-1$	β_n	β_r	Pfeifer's record values
1.7.	$\beta_rk_r-\beta_{r+1}k_{r+1}-1$	β_nk_n	β_rk_r	k_n–record values
1.8.	$\alpha_rk_r-\alpha_{r+1}k_{r+1}-1$	α_nk_n	α_rk_r	ordering via truncation
1.9.	0, if $r\ne r_1$ n_1, if $r=r_1$	$\nu-n_1-n+1$	$\nu-r+1$, if $r\le r_1$ $\nu-n_1-r+1$, if $r>r_1$	progressive type II censoring

We now introduce some notations which will be used throughout in the next section on distribution theory and in the following chapters.

NOTATIONS AND REMARKS 2.10.

i) The constant c_{r-1} is defined by

$$c_{r-1} = \prod_{i=1}^{r} \gamma_i \, , \quad r = 1,\dots,n \, ,$$

with $\gamma_n = k$.

Hence,
$$c_{n-1} = \prod_{i=1}^{n} \gamma_i = k \prod_{i=1}^{n-1} \gamma_i$$

is the constant appearing in the joint density function of uniform g OS′s (see 2.1.).

In the notation c_{r-1} the constants k, n and $\tilde{m}$ are suppressed for brevity.

If constants with different values n_1 and n_2 of n are needed, this is indicated by $c_{r-1}(n_1)$, $c_{r-1}(n_2)$.

ii) On the unit interval the functions h_m and g_m, $m \in \mathbb{R}$, are defined by

$$h_m(x) = \begin{cases} -\frac{1}{m+1}(1-x)^{m+1} \, , & m \neq -1 \\ \log\left(\frac{1}{1-x}\right) \, , & m = -1 \end{cases} \, , \quad x \in [0\, ,\, 1) \, ,$$

$$g_m(x) = h_m(x) - h_m(0) = \begin{cases} \frac{1}{m+1}\left(1-(1-x)^{m+1}\right) , & m \neq -1 \\ \log\left(\frac{1}{1-x}\right) & , m = -1 \end{cases} \, , \quad x \in [0\, ,\, 1).$$

Since the joint density function of uniform g OS′s equals

$$f^{U(1,n,\tilde{m},k),\dots,U(n,n,\tilde{m},k)}(u_1,\dots,u_n) = c_{n-1}\left(\prod_{i=1}^{n-1} \frac{d}{du_i} h_{m_i}(u_i)\right)(1-u_n)^{k-1} \, ,$$

the functions h_m and g_m frequently appear in the sequel.

This particular choice of h_m as the foundation of the above joint density function is one reason to speak of a̲ concept of g OS′s.

3. Distribution Theory of Generalized Order Statistics

Proceeding from the joint density function of the uniform g OS′s $U(r,n,\tilde{m},k)$, $r = 1,...,n$ (see Definition 2.1.), we derive the joint density of the first $r \in \{1,...,n\}$ uniform g OS′s and the one– and two–dimensional marginal densities.
By means of density transformation, we obtain the corresponding densities of g OS′s which are based on an absolutely continuous distribution function.
Then special choices of the parameters lead to representations of densities with respect to the models of ordered r.v.′s shown in Section 1.

The marginal distribution of a single uniform g OS is the basis for the results obtained in the sequel. Thus we give representations of the marginal density functions, of the marginal distribution functions and of the difference of distribution functions of successive (uniform) g OS′s. Moreover, we show some properties, e.g. recurrence relations for marginal density functions and distribution functions.

3.1. Marginal Density Functions

Starting from the joint density function of the uniform g OS′s $U(j,n,\tilde{m},k)$, $j = 1,...,n$, the marginal density of the first $r \in \{1,...,n\}$ g OS′s and by this the one– and two–dimensional marginal densities are derived.

Lemma 3.1.1. The joint density function of the first r uniform g OS′s

$$U(1,n,\tilde{m},k)\ , \ \dots\ , \ U(r,n,\tilde{m},k)\ , \quad r \in \{1,\dots,n\}\ ,$$

is given by

$$f^{U(1,n,\tilde{m},k),\dots,U(r,n,\tilde{m},k)}(u_1,\dots,u_r)$$

$$= c_{r-1} \left(\prod_{i=1}^{r-1} (1-u_i)^{m_i} \right) (1-u_r)^{k+n-r+M_r-1} \ , \quad 0 \le u_1 \le \dots \le u_r < 1\ .$$

PROOF The expression is obtained via

$$f^{U(1,n,\tilde{m},k),\dots,U(r,n,\tilde{m},k)}(u_1,\dots,u_r)$$

$$= \int_{u_r}^{1} \dots \int_{u_{n-1}}^{1} f^{U(1,n,\tilde{m},k),\dots,U(n,n,\tilde{m},k)}(u_1,\dots,u_n)\, du_n \dots du_{r+1}$$

and induction starting with $r = n-1$.

EXAMPLES

The joint density of the first r g OS's based on an absolutely continuous distribution function F with density function f is then given by

$$f^{X(1,n,\tilde{m},k),\dots,X(r,n,\tilde{m},k)}(x_1,\dots,x_r)$$

$$= c_{r-1} \left(\prod_{i=1}^{r-1} (1-F(x_i))^{m_i} f(x_i) \right) (1-F(x_r))^{k+n-r+M_r-1} f(x_r),$$

$$F^{-1}(0) < x_1 \le x_2 \le \dots \le x_r < F^{-1}(1).$$

In the case of o OS's, OS's with non–integral sample size, sequential OS's, k–records and Pfeifer's records we obtain the expressions

i) $m_1 = \dots = m_{n-1} = 0$, $k = 1$ (cf. 1.1.):

$$f^{X(1,n,0,1),\dots,X(r,n,0,1)}(x_1,\dots,x_r) = \frac{n!}{(n-r)!} \left(\prod_{i=1}^{r-1} f(x_i) \right) (1-F(x_r))^{n-r} f(x_r),$$

ii) $m_1 = \dots = m_{n-1} = 0$, $k = \alpha - n + 1$ (cf. 1.2.):

$$f^{X(1,n,0,\alpha-n+1),\dots,X(r,n,0,\alpha-n+1)}(x_1,\dots,x_r)$$

$$= \left(\prod_{j=1}^{r} (\alpha-j+1) \right) \left(\prod_{i=1}^{r-1} f(x_i) \right) (1-F(x_r))^{\alpha-r} f(x_r),$$

iii) $m_i = (n-i+1)\,\alpha_i - (n-i)\,\alpha_{i+1} - 1$, $1 \le i \le n-1$, $k = \alpha_n$ (cf. 1.3.):

$$f^{X(1,n,\tilde{m},\alpha_n),\dots,X(r,n,\tilde{m},\alpha_n)}(x_1,\dots,x_r)$$

$$= \frac{n!}{(n-r)!} \left(\prod_{j=1}^{r} \alpha_j \right) \left(\prod_{i=1}^{r-1} (1-F(x_i))^{m_i} f(x_i) \right) (1-F(x_r))^{\alpha_r(n-r+1)-1} f(x_r),$$

iv) $m_1 = \dots = m_{n-1} = -1$, $k \in \mathbb{N}$ (cf. 1.5.) :

$$f^{X(1,n,-1,k),\dots,X(r,n,-1,k)}(x_1,\dots,x_r) = k^r \left(\prod_{i=1}^{r-1} \frac{f(x_i)}{1 - F(x_i)} \right) (1 - F(x_r))^{k-1} f(x_r) ,$$

v) $m_i = \beta_i - \beta_{i+1} - 1$, $1 \le i \le n-1$, $k = \beta_n$ (cf. 1.6.) :

$$f^{X(1,n,\tilde{m},\beta_n),\dots,X(r,n,\tilde{m},\beta_n)}(x_1,\dots,x_r)$$

$$= \left(\prod_{j=1}^{r} \beta_j \right) \left(\prod_{i=1}^{r-1} (1 - F(x_i))^{m_i} f(x_i) \right) (1 - F(x_r))^{\beta_r - 1} f(x_r) .$$

In the derivation of the one–dimensional marginal densities and distribution functions the choice of the vector $\tilde{m}$ is restricted (see Remark 3.1.3.). Simple and useful expressions result, if the assumption $m_1 = \dots = m_{r-1}$ is made with respect to the r–th g OS $X(r,n,\tilde{m},k)$.

LEMMA 3.1.2. The marginal density function of the uniform g OS $U(r,n,\tilde{m},k)$, $r \in \{1,\dots,n\}$, $m_1 = \dots = m_{r-1} = m$, say, is given by

$$f^{U(r,n,\tilde{m},k)}(u) = \frac{c_{r-1}}{(r-1)!} (1-u)^{k+n-r+M_r-1} g_m^{r-1}(u) .$$

PROOF The case $r = 1$ is contained in Lemma 3.1.1. Let $r \ge 2$.
First, the auxiliary result

$$A_j = \int_{u_{r-j-1}}^{u_r} \dots \int_{u_{r-2}}^{u_r} \prod_{i=1}^{r-1} h_m'(u_i) \, du_{r-1} \dots du_{r-j}$$

$$= \left(\prod_{i=1}^{r-j-1} h_m'(u_i) \right) \frac{1}{j!} \left(h_m(u_r) - h_m(u_{r-j-1}) \right)^j , \; j = 1,2,\dots ,$$

is shown by induction on j .

Let $j = 1$:

$$\int_{u_{r-2}}^{u_r} \prod_{i=1}^{r-1} h_m'(u_i)\, du_{r-1} = \prod_{i=1}^{r-2} h_m'(u_i) \int_{u_{r-2}}^{u_r} (1-u_{r-1})^m\, du_{r-1}$$

$$= \prod_{i=1}^{r-2} h_m'(u_i)\,(h_m(u_r) - h_m(u_{r-2}))\,.$$

Step $j \rightarrow j+1$:

In the representation

$$A_{j+1} = \int_{u_{r-j-2}}^{u_r} \cdots \int_{u_{r-2}}^{u_r} \prod_{i=1}^{r-1} h_m'(u_i)\, du_{r-1} \ldots du_{r-j-1} = \left(\prod_{i=1}^{r-j-2} h_m'(u_i) \right) \tfrac{1}{j!}\, I$$

the integral I has to be evaluated :

$$I = \int_{u_{r-j-2}}^{u_r} \left(h_m(u_r) - h_m(u_{r-j-1}) \right)^j (1-u_{r-j-1})^m\, du_{r-j-1}$$

$$= \sum_{i=0}^{j} (-1)^i \tbinom{j}{i} \left(\int_{u_{r-j-2}}^{u_r} (1-u)^m\, h_m^i(u)\, du \right) h_m^{j-i}(u_r)$$

$$= h_m^{j+1}(u_r) \sum_{i=0}^{j} (-1)^i \tbinom{j}{i} \tfrac{1}{i+1} - \sum_{i=1}^{j+1} (-1)^{i-1} \tbinom{j}{i-1} \tfrac{1}{i}\, h_m^i(u_{r-j-2})\, h_m^{j-i+1}(u_r)$$

$$= \tfrac{1}{j+1} \left(h_m(u_r) - h_m(u_{r-j-2}) \right)^{j+1}$$

applying Corollary A.2.3.

Putting $j = r-1$ and $u_0 = 0$ the assertion follows from 3.1.1. :

$$f^{U(r,n,\tilde{m},k)}(u_r) = c_{r-1}\,(1-u_r)^{k+n-r+M_r-1} \int_0^{u_r} \cdots \int_{u_{r-2}}^{u_r} \prod_{i=1}^{r-1} h_m'(u_i)\, du_{r-1} \ldots du_1\,.$$

EXAMPLES

The marginal density of the r–th g OS′s based on an absolutely continuous distribution function F with density function f is then given by

$$f^{X(r,n,\tilde{m},k)}(x) = \tfrac{c_{r-1}}{(r-1)!}\,(1-F(x))^{k+n-r+M_r-1}\, f(x)\; g_m^{r-1}(F(x))\,.$$

In the case of o OS′s, OS′s with non–integral sample size, sequential OS′s, k–records and Pfeifer′s records we obtain the expressions

i) $m_1 = \dots = m_{n-1} = 0$, $k = 1$ (cf. 1.1.) :

$$f^{X(r,n,0,1)}(x) = r\binom{n}{r} F^{r-1}(x)\,(1-F(x))^{n-r} f(x)\,,$$

ii) $m_1 = \dots = m_{n-1} = 0$, $k = \alpha - n + 1$ (cf. 1.2.) :

$$f^{X(r,n,0,\alpha-n+1)}(x) = r\binom{\alpha}{r} F^{r-1}(x)\,(1-F(x))^{\alpha-r} f(x)\,,$$

iii) $m_i = (n-i+1)\,\alpha_i - (n-i)\,\alpha_{i+1} - 1$, $1 \le i \le n-1$, $m_1 = \dots = m_{r-1} = m$, $k = \alpha_n$ (cf. 1.3.) :

$$f^{X(r,n,\tilde{m},\alpha_n)}(x) = r\binom{n}{r}\Big(\prod_{j=1}^{r} \alpha_j\Big)(1-F(x))^{\alpha_r(n-r+1)-1} f(x)\, g_m^{r-1}(F(x))\,,$$

iv) $m_1 = \dots = m_{n-1} = -1$, $k \in \mathbb{N}$ (cf. 1.5.) :

$$f^{X(r,n,-1,k)}(x) = \frac{k^r}{(r-1)!}\Big(\log \frac{1}{1-F(x)}\Big)^{r-1} (1-F(x))^{k-1} f(x)\,,$$

v) $m_i = \beta_i - \beta_{i+1} - 1$, $1 \le i \le n-1$, $k = \beta_n$ (cf. 1.6.) :

$$f^{X(r,n,\tilde{m},\beta_n)}(x) = \frac{1}{(r-1)!}\Big(\prod_{j=1}^{r} \beta_j\Big)(1-F(x))^{\beta_r - 1} f(x)\, g_m^{r-1}(F(x))\,.$$

REMARK 3.1.3. The simplicity of the representation is based on identical constants m_i, $i = 1,\dots,r-1$; otherwise, e.g. for $r = n$ and $j = 2$, the expression

$$\tfrac{1}{2}\Big(h_m(u_n) - h_m(u_{n-3}) \Big)^2$$

is replaced by

$$\frac{m_{n-1}+1}{m_{n-1}+m_{n-2}+2}\, h_{m_{n-1}}(u_n)\, h_{m_{n-2}}(u_n) - h_{m_{n-1}}(u_n)\, h_{m_{n-2}}(u_{n-3})$$
$$+ \frac{m_{n-2}+1}{m_{n-1}+m_{n-2}+2}\, h_{m_{n-1}}(u_{n-3})\, h_{m_{n-2}}(u_{n-3})$$

in the auxiliary result contained in the proof of 3.1.2.

However, a representation of the 'generalized minimum' $X(1,n,\tilde{m},k)$ is included in the representation of the joint density of the first r g OS's for arbitrary $m_1,\dots,m_{n-1}$ (see Remark 3.1.5.).

Considering the marginal density of $X(r,n,\tilde{m},k)$, we observe identical distributions of g OS's with some negative parameter m and those with $m' = -(m+2)$ in the case $m_1 = \dots = m_{n-1} = m$, $m < -1$.

REMARK 3.1.4. The g OS's $X(r,n,m,k)$ and $X(r,n',-m-2,k')$ with

$$k',n' \in \mathbb{N}\,,\; n' \geq r\,,\; m < -1 \;\text{ and }\; k - k' = -(n + n' - r - 1)(m + 1)$$

(i.e. $k' \leq k$) have identical densities :

$$f^{X(r,n,m,k)}(x) = f^{X(r,n',-m-2,k')}(x)\,.$$

In particular, choosing $m = -2$, $k' = 1$, $r = n$ and $n' = k$ ($\geq n$), the distributions of a certain g OS and some ordinary maximum ($m + 2 = 0$) in a sample of size k ($k \geq n$) are identical.

REMARK 3.1.5. It is also clear from the representation of the marginal density in the case $r = 1$ that the minimum $X(1,k,0,1)$ of k random variables with distribution function F and the first k–th record $X(1,n,-1,k)$ based on F have identical distributions (see Remark 3.3.3.).

More generally we find :

$X(1,n,\tilde{m},k)$, $X(1,n,-1,k+n-1+M_1)$ and $X(1,k+n-1+M_1,0,1)$ are identically distributed;

i.e. any first g OS can be viewed as some minimum or as some first record.

(The function g_m drops out if $r = 1$.)

REMARK 3.1.6. The representation of the marginal density of $U(r,n,m,k)$ yields identical distributions of the uniform g OS's $U(r,n,m,k+m+1)$ and $U(r,n+1,m,k)$:

Putting $k' = k + m + 1$ we have

$$f^{U(r,n,m,k')}(u) = \frac{c_{r-1}(k')}{(r-1)!}(1-u)^{k'+(n-r)(m+1)-1}\, g_m^{\,r-1}(u)$$

$$= \frac{c_{r-1}(n+1)}{(r-1)!}(1-u)^{k+(n-r+1)(m+1)-1}\, g_m^{\,r-1}(u)$$

$$= f^{U(r,n+1,m,k)}(u)\,.$$

The assertion is trivial in the case of records ($m = -1$) .

It may be noted already that there are recurrence relations and identities for one–dimensional marginal density functions of g OS's (see Lemma 3.2.2. and 3.2.3.).

For the two–dimensional marginal densities the choice of $\tilde{m}$ is again restricted to obtain useful expressions.

LEMMA 3.1.7. The joint density functions of the uniform g OS's $U(r,n,\tilde{m},k)$ and $U(s,n,\tilde{m},k)$, $r < s$, with

$$m_1 = \dots = m_{r-1} = \mu,\ m_{r+1} = \dots = m_{s-1} = \nu$$

(no assumption, if $s - r = 1$), is given by ($u_r < u_s$)

$$f^{U(r,n,\tilde{m},k),\, U(s,n,\tilde{m},k)}(u_r,u_s)$$

$$= \frac{c_{s-1}}{(r-1)!\,(s-r-1)!}\,(1-u_r)^{m_r}\, g_\mu^{r-1}(u_r) \cdot \Big(h_\nu(u_s) - h_\nu(u_r)\Big)^{s-r-1} (1-u_s)^{k+n-s+M_s-1} .$$

PROOF Let $s - r \geq 2$; the assertion remains valid for $s - r = 1$.
Starting point is the joint density of the first s uniform g OS's. Thus, applying the auxiliary result in the proof of Lemma 3.1.2. twice, we obtain:

$$f^{U(r,n,\tilde{m},k),\, U(s,n,\tilde{m},k)}(u_r,u_s)$$

$$= \int_0^{u_r}\!\!\cdots\int_{u_{r-2}}^{u_r}\int_{u_r}^{u_s}\!\!\cdots\int_{u_{s-2}}^{u_s} f^{U(1,n,\tilde{m},k),\dots,U(s,n,\tilde{m},k)}(u_1,\dots,u_s)\, du_{s-1}\dots du_{r+1}\, du_{r-1}\dots du_1$$

$$= c_{s-1}\,(1-u_r)^{k+n-s+M_s-1} \int_0^{u_r}\!\!\cdots\int_{u_{r-2}}^{u_r}\Big(\prod_{i=1}^{r-1} h'_\mu(u_i)\Big)$$

$$\cdot\,(1-u_r)^{m_r}\int_{u_r}^{u_s}\!\!\cdots\int_{u_{s-2}}^{u_s}\Big(\prod_{i=r+1}^{s-1} h'_\nu(u_i)\Big)\, du_{s-1}\dots du_{r+1}\, du_{r-1}\dots du_1$$

$$= c_{s-1}\,(1-u_r)^{m_r}(1-u_s)^{k+n-s+M_s-1}\cdot\frac{1}{(s-r-1)!}\Big(h_\nu(u_s)-h_\nu(u_r)\Big)^{s-r-1}$$

$$\cdot\int_0^{u_r}\!\!\cdots\int_{u_{r-2}}^{u_r}\Big(\prod_{i=1}^{r-1} h'_\mu(u_i)\Big)\, du_{r-1}\dots du_1$$

$$= \frac{c_{s-1}}{(s-r-1)!}(1-u_r)^{m_r}\Big(h_\nu(u_s)-h_\nu(u_r)\Big)^{s-r-1}\cdot(1-u_s)^{k+n-s+M_s-1}\,\frac{1}{(r-1)!}\, g_\mu^{r-1}(u_r) .$$

EXAMPLES The joint density functions of the g OS's $X(r,n,\tilde{m},k)$ and $X(s,n,\tilde{m},k)$, $r < s$, based on an absolutely continuous distribution function F with density function f is then given by

$$f^{X(r,n,\tilde{m},k),\, X(s,n,\tilde{m},k)}(x_r,x_s)$$

$$= \frac{c_{s-1}}{(r-1)!\,(s-r-1)!}\,(1-F(x_r))^{m_r} f(x_r)\, g_\mu^{r-1}(F(x_r))$$

$$\cdot \left(h_\nu(F(x_s)) - h_\nu(F(x_r))\right)^{s-r-1} (1-F(x_s))^{k+n-s+M_s-1} f(x_s)\,,\quad x_r < x_s\,.$$

In the case of o OS's and k–th records we obtain the expressions

i) $m = 0\,,\ k = 1$:

$$f^{X(r,n,0,1),\, X(s,n,0,1)}(x_r,x_s)$$

$$= \frac{n!}{(n-s)!(r-1)!(s-r-1)!}\, F^{r-1}(x_r)\, f(x_r)\, (F(x_s) - F(x_r))^{s-r-1}\, (1-F(x_s))^{n-s} f(x_s)\,,$$

ii) $m = -1\,,\ k \in \mathbb{N}$:

$$f^{X(r,n,-1,k),\, X(s,n,-1,k)}(x_r,x_s)$$

$$= \frac{k^s}{(r-1)!\ (s-r-1)!}\ \frac{f(x_r)}{1-F(x_r)} \left(\log\frac{1}{1-F(x_r)}\right)^{r-1}$$

$$\cdot \left(\log\frac{1}{1-F(x_s)} - \log\frac{1}{1-F(x_r)}\right)^{s-r-1} (1-F(x_s))^{k-1} f(x_s)$$

(e.g. David 1981, Nayak 1981).

REMARKS 3.1.8.

i) The expression in 3.1.7. simplifies in the case of successive uniform g OS's, since the term $\left(h_\nu(u_s) - h_\nu(u_r)\right)^{s-r-1}$ in the representation of the two–dimensional marginal density drops out.
In particular, if $r = 1$ and $s = n$, then $m_2 = \ldots = m_{n-1} = \nu$ has to be chosen and we find

$$f^{U(1,n,\tilde{m},k),\, U(n,n,\tilde{m},k)}(u_1,u_n)$$
$$= \frac{c_{n-1}}{(n-2)!}(1-u_1)^{m_1}\left(h_\nu(u_n)-h_\nu(u_1)\right)^{n-2}(1-u_n)^{k-1}.$$

ii) Analogously, expressions for any s–dimensional marginal density

$$f^{U(r_1,n,m,k),\dots,U(r_s,n,m,k)}(u_{r_1},\dots,u_{r_s}),\ r_1<\dots<r_s \text{ and } \{r_1,\dots,r_s\}\subset\{1,\dots,n\}$$

may also be obtained (see David 1981, p 10 for o OS's).

Now, the density of a contrast $X(s,n,\tilde{m},k) - X(r,n,\tilde{m},k) = W_{rs}$ of g OS's can be derived (see David 1981, p 11 for o OS's).

Density transformation yields ($w = x_s - x_r$)

$$f^{W_{rs}}(w) = \frac{c_{s-1}}{(r-1)!\,(s-r-1)!}\int_{-\infty}^{\infty}(1-F(x_r))^{m_r} f(x_r)\, g_\mu^{r-1}(F(x_r))$$
$$\cdot\left(h_\nu(F(w+x_r))-h_\nu(F(x_r))\right)^{s-r-1}(1-F(w+x_r))^{k+n-s+M_s-1} f(w+x_r)\,dx_r$$

and, in particular,

$$f^{W_{1n}}(w) = \frac{c_{n-1}}{(n-2)!}\int_{-\infty}^{\infty}(1-F(x_1))^{m_1} f(x_1)$$
$$\cdot\left(h_\nu(F(w+x_1))-h_\nu(F(x_1))\right)^{n-2}(1-F(w+x_1))^{k-1} f(w+x_1)\,dx_1\,.$$

for the range (i.e. $r = 1$, $s = n$).

Thus, the distribution function of the range W_{1n} is given by (see David 1981, p 12 for o OS's)

$$F^{W_{1n}}(w) = \frac{c_{n-1}}{(n-2)!}\int_{-\infty}^{\infty}(1-F(x))^{m_1} f(x)$$
$$\cdot\int_{-\infty}^{w}\left(h_\nu(F(t+x))-h_\nu(F(x))\right)^{n-2}(1-F(t+x))^{k-1} f(t+x)\,dt\,dx$$

where the inner integral may be solved via iterated integration by parts.

3.2. Properties and Distribution Function of a Single Generalized Order Statistic

Since g OS's based on some absolutely continuous distribution function F are defined via the quantile transformation of uniform g OS's, the density function of the r–th g OS $X(r,n,\tilde{m},k)$, $m_1 = \dots = m_{r-1} = m$, is expressible as (see 2.3.)

$$f^{X(r,n,\tilde{m},k)}(x) = f^{U(r,n,\tilde{m},k)}(F(x))\, f(x)\,.$$

Thus, the marginal densities of uniform g OS's play an important role in the investigation of g OS's. We introduce a short notation and show some properties.

NOTATION 3.2.1. Let $m_1 = \dots = m_{r-1} = m$.
The density function of the r–th uniform g OS is denoted by

$$\varphi_{r,n}(x) = f^{U(r,n,\tilde{m},k)}(x) = \frac{c_{r-1}}{(r-1)!}(1-x)^{\gamma_r - 1}\, g_m^{\,r-1}(x)\,,\quad x \in (0,1)\,,\ 1 \le r \le n\,,$$

(see Lemma 3.1.2.).

In this notation the parameters $\tilde{m}$ and k are suppressed for brevity.

EXAMPLES

The density of the r–th o OS

$$f^{X(r,n,0,1)}(x) = r\binom{n}{r} F^{r-1}(x)\,(1-F(x))^{n-r} f(x)$$

is expressible as
$$f^{X(r,n,0,1)}(x) = \varphi_{r,n}(F(x))\, f(x)$$

where
$$\varphi_{r,n}(x) = r\binom{n}{r} x^{r-1}(1-x)^{n-r}\,,\ x \in (0,1)\,,$$

is the density function of a beta–distributed random variable with parameters r and $n-r+1$ (see David 1981, p 8/9, Blom 1958, Part II). Hence, properties of a beta–density or the beta–integral can be utilized.

The density of the k–th record value ($m = -1$, $k \in \mathbb{N}$) is given by

$$f^{X(r,n,-1,k)}(x) = \varphi_{r,n}(F(x))\, f(x)$$

with
$$\varphi_{r,n}(x) = \frac{k^r}{(r-1)!}\left(\log \frac{1}{1-x}\right)^{r-1} (1-x)^{k-1} .$$

Thus, $\varphi_{r,n}$ is the density of the random variable

$$X = 1 - e^{-Y}$$

where Y possesses a gamma distribution with parameters k and r and density function

$$f^{Y}(y) = \frac{k^r}{(r-1)!}\, y^{r-1}\, e^{-ky} , \; y > 0 .$$

It is of interest to notice that the density $\varphi_{r,n}$ (for $m = -1$ and $k \in \mathbb{N}$) arises in another context (see Rider 1955). There (see also Rahman 1964, David 1981) $\varphi_{r,n}(1-x)$ turns out to be the density of a product of r maxima from independent samples of size k with an underlying standard uniform distribution.

Recurrence relations for $\varphi_{r,n}$ are now presented, leading also to identities for one–dimensional density (and distribution) functions of g OS's.

Lemma 3.2.2. In the case $m_1 = \ldots = m_{n-1} = m$ the following recurrence relations hold for densities as in 3.2.1. with $1 \le r \le n-1$:

i)
$$(k+(n-r-1)(m+1))\,\varphi_{r,n}(x) + r(m+1)\,\varphi_{r+1,n}(x) = (k+(n-1)(m+1))\,\varphi_{r,n-1}(x) ,$$

ii)
$$(k+(n-r-1)(m+1))\,(\varphi_{r+1,n}(x) - \varphi_{r,n}(x)) = (k+(n-1)(m+1))\,(\varphi_{r+1,n}(x) - \varphi_{r,n-1}(x)) ,$$

iii)
$$(k+(n-1)(m+1))\,(\varphi_{r,n}(x) - \varphi_{r,n-1}(x)) = r(m+1)\,(\varphi_{r,n}(x) - \varphi_{r+1,n}(x)) .$$

Proof i) If $m = -1$, then $\varphi_{r,n}(x) = \varphi_{r,n-1}(x)$ and the equation is trivial. Let $m \neq -1$. Noticing that

$$c_r(n) = (k+(n-r-1)(m+1))\, c_{r-1}(n) \quad\text{and}\quad c_r(n) = (k+(n-1)(m+1))\, c_{r-1}(n-1)\,,$$

we have

$$(k+(n-r-1)(m+1))\, \varphi_{r,n}(x) + r(m+1)\, \varphi_{r+1,n}(x)$$

$$= \frac{c_r(n)}{(r-1)!}(1-x)^{k+(n-r-1)(m+1)-1}\, g_m^{r-1}(x)\left[(1-x)^{m+1} + (m+1)\, g_m(x)\right]$$

$$= (k+(n-1)(m+1))\, \varphi_{r,n-1}(x)\,.$$

ii) and iii) are directly obtained from i).

In the case of record values ($m = -1$) the identities are trivial. But for $m \neq -1$, 3.2.2.i) is a generalization of a well-known identity for densities of o OS's; i.e. of a recurrence relation for the incomplete beta-function (e.g. David 1981, p 46/7).

COROLLARY 3.2.3. Consider g OS's based on an absolutely continuous distribution function F with density function f.

Since $$f^{X(r,n,\tilde{m},k)}(x) = \varphi_{r,n}(F(x))\, f(x)\,,$$

the above relations are valid for densities of g OS's.

To prepare for a representation of the marginal distribution function of a g OS $X(r,n,\tilde{m},k)$, we state the distribution function $\Phi_{r,n}$ corresponding to $\varphi_{r,n}$.

LEMMA 3.2.4. The distribution function $\Phi_{r,n}$ corresponding to the density function $\varphi_{r,n}$ (see 3.2.1.), $m_1 = \ldots = m_{r-1} = m$ and $r \in \{1,\ldots,n\}$, is given by

$$\Phi_{r,n}(x) = 1 - c_{r-1}\,(1-x)^{k+n-r+M_r} \sum_{j=0}^{r-1} \frac{1}{j!\; c_{r-j-1}}\, g_m^{j}(x)\,,\quad x \in (0,1)\,.$$

COROLLARY 3.2.5. Since $$\Phi_{r,n}(x) = \int_0^x \varphi_{r,n}(t)\, dt\,,$$

the recurrence relations in Lemma 3.2.2. are also valid for the distribution functions of uniform g OS's ($m_1 = \ldots = m_{n-1} = m$, $1 \le r \le n-1$).

Using the quantile transformation (cf. 2.3.), we obtain a representation of the distribution function of a g OS based on an arbitrary distribution function F (see 3.2.4.) :

$$F^{X(r,n,\tilde{m},k)}(x) = \Phi_{r,n}(F(x)) .$$

COROLLARY 3.2.6. Let $m_1 = \ldots = m_{r-1} = m$ and let F be an arbitrary distribution function. Then the marginal distribution function of the r–th g OS $X(r,n,\tilde{m},k)$, $r \in \{1,\ldots,n\}$ (based on F, see 2.3.) is given by

$$F^{X(r,n,\tilde{m},k)}(x) = \Phi_{r,n}(F(x)) = 1 - c_{r-1}\,(1-F(x))^{k+n-r+M_r} \sum_{j=0}^{r-1} \frac{1}{j!\;c_{r-j-1}}\, g_m^{\,j}(F(x)) .$$

EXAMPLES In the case of o OS's, OS's with non–integral sample size, k–th record values and Pfeifer's records the following expressions result :

i) $m = 0,\ k = 1$:

$$F^{X(r,n,0,1)}(x) = 1-(1-F(x))^{n-r+1} \sum_{j=0}^{r-1} \binom{n-r+j}{j} F^j(x)$$

$$= 1 - \sum_{j=0}^{r-1} \binom{n}{j} F^j(x)\,(1-F(x))^{n-j}$$

(applying Lemma A.2.1.; e.g. David 1981),

ii) $m = 0,\ k = \alpha - n + 1$:

$$F^{X(r,n,0,\alpha-n+1)}(x) = 1 - \sum_{j=0}^{r-1} \binom{\alpha}{j} F^j(x)\,(1-F(x))^{\alpha-j}$$

(cf. Rohatgi, Saleh 1988),

iii) $m = -1,\ k \in \mathbb{N}$:

$$F^{X(r,n,-1,k)}(x) = 1-(1-F(x))^k \sum_{j=0}^{r-1} \frac{1}{j!}\Big(k \log \frac{1}{1-F(x)}\Big)$$

(cf. Grudzień, Szynal 1983),

iv) $m_i = \beta_i - \beta_{i+1} - 1,\ 1 \le i \le n-1,\ k = \beta_n$:

$$F^{X(r,n,\tilde{m},\beta_n)}(x) = 1-(1-F(x))^{\beta_r} \sum_{j=0}^{r-1} \frac{1}{j!}\Big(\prod_{i=r-j+1}^{r} \beta_i\Big)\, g_m^{\,j}(F(x)) .$$

REMARK 3.2.7. Applying A.2.2., the distribution function in 3.2.6. may be written as

$$F^{X(r,n,\tilde{m},k)}(x)$$
$$= 1 - \frac{c_{r-1}}{(m+1)^{r-1}} \sum_{j=0}^{r-1} (-1)^j \frac{1}{j!\,(r-j-1)!} \frac{1}{k+n-r+M_r+j(m+1)} (1-F(x))^{k+n-r+M_r+j(m+1)},$$

if $m \neq -1$.

In particular, in the case of o OS's we find

$$F^{X(r,n,0,1)}(x) = 1 - \frac{n!}{(n-r)!} \sum_{j=0}^{r-1} (-1)^{r-j-1} \frac{1}{j!\,(r-j-1)!} \frac{1}{n-j} (1-F(x))^{n-j}$$

(cf. Lemma A.2.4. with $F(x) = x$).

Using 3.2.4., an expression for the difference $\Phi_{r,n}(x) - \Phi_{r-1,n}(x)$ is obtained which is shown in 3.2.10. for distribution functions of g OS's and which is important in connection with recurrence relations for moments of g OS's (see Section III.2.).

COROLLARY 3.2.8. Given the distribution functions $\Phi_{r,n}$ and $\Phi_{r-1,n}$ (see Lemma 3.2.4.) with $m_1 = \dots = m_{r-1} = m$, we have for $r \geq 2$:

$$\Phi_{r,n}(x) - \Phi_{r-1,n}(x) = -\frac{c_{r-2}}{(r-1)!} (1-x)^{k+n-r+M_r} g_m^{\,r-1}(x), \quad x \in (0,1).$$

PROOF The assertion may be seen by direct calculation or via integration by parts. Noticing that $g_m(0) = 0$,

$$\Phi_{r,n}(x) = \int_0^x \varphi_{r,n}(t)\,dt = -\frac{c_{r-2}}{(r-1)!}(1-t)^{k+n-r+M_r} g_m^{\,r-1}(t)\Big|_0^x$$
$$+ \frac{c_{r-2}}{(r-2)!} \int_0^x (1-t)^{k+n-r+M_r} g_m^{\,r-2}(t)\, g_m'(t)\,dt$$
$$= -\frac{c_{r-2}}{(r-1)!}(1-x)^{k+n-r+M_r} g_m^{\,r-1}(x) + \Phi_{r-1,n}(x).$$

REMARK 3.2.9. Together with 3.2.2., 3.2.5. and 3.2.8., there are also representations available of

$$\Phi_{r,n}(x) - \Phi_{r-1,n-1}(x) \text{ and } \Phi_{r-1,n}(x) - \Phi_{r-1,n-1}(x)$$

$(m_1 = \ldots = m_{n-1} = m)$.

In this connection, we refer to Corollary 3.2.10. for distribution functions and to Corollary III.1.1. for moments of g OS's.

The identities in 3.2.5. (3.2.2.) and 3.2.8. for functions $\Phi_{r,n}$ (see 3.2.4.) lead to useful relations for distribution functions of g OS's.

COROLLARY 3.2.10. Let the cited g OS's be well defined.

i) For $r \geq 2$ and $m_1 = \ldots = m_{r-1} = m$ we have :

$$F^{X(r,n,\tilde{m},k)}(x) - F^{X(r-1,n,\tilde{m},k)}(x)$$

$$= -\frac{c_{r-2}(n)}{(r-1)!}(1 - F(x))^{k+n-r+M_r}\, g_m^{r-1}(F(x)) .$$

ii) For $r \geq 2$ and $m_1 = \ldots = m_{n-1} = m$ the following relations hold :

$$(k+(n-1)(m+1))\, F^{X(r-1,n-1,m,k)}(x)$$

$$= (k+(n-r)(m+1))\, F^{X(r-1,n,m,k)}(x) + (r-1)(m+1)\, F^{X(r,n,m,k)}(x) ,$$

$$F^{X(r,n,m,k)}(x) - F^{X(r-1,n-1,m,k)}(x)$$

$$= -\frac{c_{r-2}(n-1)}{(r-1)!}(1 - F(x))^{k+(n-r)(m+1)}\, g_m^{r-1}(F(x)) ,$$

$$F^{X(r-1,n,m,k)}(x) - F^{X(r-1,n-1,m,k)}(x)$$

$$= \frac{m+1}{k+(n-1)(m+1)}\,\frac{c_{r-2}(n)}{(r-2)!}(1 - F(x))^{k+(n-r)(m+1)}\, g_m^{r-1}(F(x)) .$$

EXAMPLES

i) In the case of o OS's and $2 \leq r \leq n$ the identities reduce to

$$F^{X(r,n,0,1)}(x) - F^{X(r-1,n,0,1)}(x) = -\binom{n}{r-1} F^{r-1}(x)\,(1-F(x))^{n-r+1},$$

$$F^{X(r,n,0,1)}(x) - F^{X(r-1,n-1,0,1)}(x) = -\binom{n-1}{r-1} F^{r-1}(x)\,(1-F(x))^{n-r+1},$$

$$F^{X(r-1,n,0,1)}(x) - F^{X(r-1,n-1,0,1)}(x) = \binom{n-1}{r-2} F^{r-1}(x)\,(1-F(x))^{n-r+1}.$$

The first equation is obvious on account of 3.2.6.; David, Shu (1978) note the equations and provide a representation for $EX^{\alpha}_{r-1,n} - EX^{\alpha}_{r-1,n-1}$, $\alpha \in \mathbb{N}$ (see Section II.1.3.).

ii) The relation in 3.2.10.i) and the first equation in 3.2.10.ii) are stated in Rohatgi, Saleh (1988) in the case of OS's with non–integral sample size.

The above formulae for o OS's are of interest, e.g., in reliability theory. Modeling the life–length of components in a technical system by iid random variables with distribution function F $(F^{-1}(0) \geq 0)$, the failure distribution of a r–out–of–n–system (cf. Barlow, Proschan 1975, p 59) equals the distribution function of the order statistic with indices $n-r+1$ and n. In such a system, all n components start working simultaneously and it fails, if $n-r+1$ or more components fail; i.e. r components are necessary for the system to work (failed components are exchanged without any loss of time). Thus, the formulae may be used to compare different r_{ν}–out–of–n_{ν}– systems (cf. Section 1.1.).

If a random variable T is distributed according to $\Phi_{r,n}$, then its expectation can be explicitly derived. It is used in a certain inequality for moments of g OS's (see IV.2.1.).

LEMMA 3.2.11. Let the random variable T be distributed with density $\varphi_{r,n}$ (see 3.2.1.) and $m_1 = \dots = m_{r-1} = m$.

Then its expectation ET is given by

$$ET = c_{r-1} \sum_{j=0}^{r-1} \frac{1}{c_{r-j-1}} \left(\prod_{i=0}^{j} (k + n - r + M_r + 1 + i(m+1)) \right)^{-1}.$$

PROOF The expression directly follows from A.1.2. and

$$ET = \int_0^1 (1 - \Phi_{r,n}(t))\, dt = c_{r-1} \sum_{j=0}^{r-1} \frac{1}{j!\, c_{r-j-1}} \int_0^1 (1-t)^{k+n-r+M_r}\, g_m^j(t)\, dt .$$

Examples In the case of o OS's and k–th records we obtain

i) $m = 0, k = 1$: $ET = \frac{r}{n+1}$, and

ii) $m = -1, k \in \mathbb{N}$: $ET = 1 - 2^{-r}$.

Other properties of the density function $\varphi_{r,n}$ are useful to derive sufficient conditions for the existence of moments of g OS's (see II.1.1.4., II.1.2.1.); in particular, we are interested in the limits from the left at the point 1 depending on the parameters $r, n, \tilde{m}$ and k:

Lemma 3.2.12. Let $\varphi_{r,n}$ be a density function according to 3.2.1. with $m_1 = \ldots = m_{r-1} = m$. Then $\varphi_{r,n}(x)$, $x \in (0,1)$, possesses a finite upper bound, if the constants satisfy the following condition:

$$\{r = 1\} \vee \{r > 1 \wedge m > -1\} \vee \{r > 1 \wedge m \leq -1 \wedge k + n - r + M_r > 1\}.$$

3.3. Transformations and Properties of Generalized Order Statistics

In this section, special g OS's with identical structural properties are shown; i.e. conditional distributions, stochastical independence of certain statistics and trans–formations within the class of g OS's.
It is clear from Pfeifer's model that g OS's have Markovian structure (see 2.8.). Concerning o OS's and records, this analogy is taken up by Deheuvels (1984) and Gupta (1984); their results are implicitly contained in the dissertation of Pfeifer (1979).

Remark 3.3.1. Deheuvels (1984) shows (see also Gupta 1984), that the sequence $(X(r,n,-1,k))_{r=2,\ldots,n}$ of k–th record values and the sequence $(X(n-k+1,n,0,1))_{k=1,\ldots,n-1}$ of o OS's possess identical transition probabilities ($t \geq s$):

$$P(\,X(r,n,-1,k) > t \mid X(r-1,n,-1,k) = s\,)$$

$$= P(\,X(n-k+1,n,0,1) > t \mid X(n-k,n,0,1) = s\,) = \left(\frac{1-F(t)}{1-F(s)}\right)^k, \; 1 \leq k \leq n-1\,.$$

Moreover,

$$P(\, X(n,n,m,k) > t \mid X(n-1,n,m,k) = s\,) = \left(\frac{1 - F(t)}{1 - F(s)}\right)^k$$

and for $m \le 0$, $k \in \{1,...,n-1\}$, we have

$$P(\, X(n-k+1,n,m,1-m(k-1)) > t \mid X(n-k,n,m,1-m(k-1)) = s\,) = \left(\frac{1 - F(t)}{1 - F(s)}\right)^k .$$

More generally, we find

LEMMA 3.3.2. For any $j \in \mathbb{N}$, the Markov chains of g OS's

$$X(r,n,-1,(k-1)(m+1)+j)\;,\; 2 \le r \le n\;, \text{ and } X(n-k+1,n,m,j)\;,\; 1 \le k \le n-1\;,$$

possess the transition probabilities ($t \ge s$)

$$P(\, X(r,n,-1,(k-1)(m+1)+j) > t \mid X(r-1,n,-1,(k-1)(m+1)+j) = s\,)$$

$$= P(\, X(n-k+1,n,m,j) > t \mid X(n-k,n,m,j) = s\,) = \left(\frac{1 - F(t)}{1 - F(s)}\right)^{j+(k-1)(m+1)} .$$

REMARKS 3.3.3.

i) Putting $m = j-1$, the quantity $\left(\frac{1 - F(t)}{1 - F(s)}\right)^{jk}$ ($t \ge s$) turns out to be the transition probability of the Markov chains

$(X(r,n,-1,kj))_{2 \le r \le n}$ (kj–th records based on the distribution function F) and $(X(n-k+1,n,j-1,j))_{k=1,\dots,n-1}$ (o OS's from the distribution of the minimum of j iid random variables with distribution function F).

Deheuvels' (1984) remark corresponds to $j = 1$, $m = 0$ (see 3.3.1.).

ii) Let $\nu \in \mathbb{Z}$ such that $X(r,n,\nu,(k-1)(m+1)+j)$, $r = 1,...,n$, are well defined g OS's. Then we have (cf. 3.3.2.) ($t \ge s$) :

$$P(\, X(n,n,\nu,(k-1)(m+1)+j) > t \mid X(n-1,n,\nu,(k-1)(m+1)+j) = s\,)$$

$$= \left(\frac{1 - F(t)}{1 - F(s)}\right)^{j+(k-1)(m+1)} .$$

Within the class of g OS's there are more structural similarities concerning transition probabilities. We show another example :

EXAMPLE If the cited g OS's are well defined, then we have ($t \geq s$) :

$$P(\, X(r,n,-1,k^2) > t \mid X(r-1,n,-1,k^2) = s\,)$$

$$= P(\, X(n,n,m,k^2) > t \mid X(n-1,n,m,k^2) = s\,)$$

$$= P(\, X(n-k+1,n,k-1,k) > t \mid X(n-k,n,k-1,k) = s\,) = \left(\frac{1 - F(t)}{1 - F(s)}\right)^{k^2} .$$

As an example for the independence of certain statistics, a result of Deheuvels (1984) with respect to o OS's and record values is taken up and extended to g OS's.
First of all, Lemma 1.6.6. and 2.8. , 1.6.3. respectively, together imply that the statements

{ $X(r-1,n,-1,k)$ and $X(r,n,-1,k) - X(r-1,n,-1,k)$ are stochastically independent } and

{ $X(n-k,n,0,1)$ and $X(n-k+1,n,0,1) - X(n-k,n,0,1)$ are stochastically independent }

are equivalent for $2 \leq r \leq n$ and $1 \leq k \leq n-1$.

In the case of g OS's the analogous result reads :

LEMMA 3.3.4. Let $r_i \geq 2$, $i = 1,2$, and

$$X(r_i,n_i,m_i,k_i)\,,\; X(r_i-1,n_i,m_i,k_i)\,,\; i = 1,2\,,$$

be two pairs of g OS's, and let the condition $F(x) < 1$ for all x be fulfilled with respect to the underlying distribution function F. Then we observe :

{ $X(r_1-1,n_1,m_1,k_1)$ and $X(r_1,n_1,m_1,k_1) - X(r_1-1,n_1,m_1,k_1)$ are independent } and

{ $X(r_2-1,n_2,m_2,k_2)$ and $X(r_2,n_2,m_2,k_2) - X(r_2-1,n_2,m_2,k_2)$ are independent }

are equivalent statements.

In an analogous manner the characterization results of Pfeifer (1979, 1982a) (see Section 1.6.) can be formulated for g OS's. In particular, well known results in the case of o OS's and records are contained.

As a motivation to give a unified approach using Markov chains, Deheuvels (1984) refers to the articles of

Fisz (1958), Rossberg (1960), Ferguson (1964) and Crawford (1966) :

Given the order statistics based on an absolutely continuous distribution function F.
Then $X(1,2,0,1)$ and $X(2,2,0,1) - X(1,2,0,1)$ are independent,
iff F is the distribution function of an exponential distribution.

and of

Tata (1969) :

Given the records based on an absolutely continuous distribution function F.
Then $X(1,n,-1,1)$ and $X(2,n,-1,1) - X(1,n,-1,1)$ are independent,
iff F is the distribution function of an exponential distribution.

Using Theorem 1.6.7., an analogon for g OS's results.

Based on the standard exponential distribution ($F(x) = 1 - e^{-x}$), it is well known that the normalized spacings

$$Y_i = (n - i + 1)(X_{i,n} - X_{i-1,n}) \, , \; i = 1,\dots,n \, ,$$

with $X_{0,n} = 0$, are stochastically independent and again distributed according to F. This important result is stated in Sukhatme (1937) and Rényi (1953) (see David 1981, p 20/1). Rohatgi, Saleh (1988) point out that the same property holds true for OS's with non–integral sample size. Resnick (1973a) shows an analogous behaviour of records; i.e. the random variables

$$Z_1 = X_{L(1)} \, , \; Z_i = X_{L(i)} - X_{L(i-1)} \, , \; i \in \mathbb{N}^{\geq 2} \, ,$$

are stochastically independent and standard exponentially distributed (cf. Section 1.6., see e.g. Pfeifer 1989, p 83).

Thus, the r–th o OS $X_{r,n}$ may be represented as a weighted sum of iid random variables

$$X_{r,n} = \sum_{i=1}^{r} (X_{i,n} - X_{i-1,n}) = \sum_{i=1}^{r} \frac{Y_i}{n - i + 1}$$

and the r–th record as a sum $\qquad X_{L(r)} = \sum_{i=1}^{r} Z_i \, .$

By this, asymptotic properties of the sequences $(R(X_{L(n)}))_n$, $R(x) = -\log(1 - F(x))$ (cf. Section 1.6.) can be deduced, e.g. a central limit theorem, a strong law of large numbers and a law of the iterated logarithm (see Resnick 1973a,b).

The generalization of the assertions to g OS's is subject matter of the following theorem; g OS's can also be represented as a sum of iid random variables.

THEOREM 3.3.5. Let $X(j,n,\tilde{m},k)$, $j = 1,...,n$, be g OS's based on the distribution function F with $F(x) = 1 - e^{-x}$, $x \geq 0$. Then we find :

The random variables

$$Y_1 = \gamma_1 X(1,n,\tilde{m},k) ,$$

$$Y_j = \gamma_j(X(j,n,\tilde{m},k) - X(j-1,n,\tilde{m},k)) , j = 2,...,n ,$$

with $\quad \gamma_j = k + n - j + M_j$

are stochastically independent and identically distributed according to F.

Moreover, we have the representation

$$X(r,n,\tilde{m},k) \left(= X(1,n,\tilde{m},k) + \sum_{j=2}^{r} (X(j,n,\tilde{m},k) - X(j-1,n,\tilde{m},k)) \right) = \sum_{j=1}^{r} Y_j/\gamma_j .$$

PROOF Let $F(x) = 1 - e^{-x}$, $x \geq 0$. Then

$$f^{X(1,n,\tilde{m},k),...,X(n,n,\tilde{m},k)}(x_1,...,x_n)$$

$$= k \left(\prod_{j=1}^{n-1} \gamma_j \right) \left(\prod_{i=1}^{n-1} (1 - F(x_i))^{m_i} f(x_i) \right) (1 - F(x_n))^{k-1} f(x_n)$$

$$= k \left(\prod_{j=1}^{n-1} \gamma_j \right) \exp \left(- \sum_{j=1}^{n} \gamma_j(x_j - x_{j-1}) \right) , \; x_0 = 0 , \; x_1 \leq ... \leq x_n .$$

From the definition of Y_i and density transformation we obtain

$$f^{Y_1,...,Y_n}(y_1,...,y_n) = \prod_{j=1}^{n} \exp(-y_j) .$$

REMARK 3.3.6. If $m = k - 1$ is assumed in addition, then

$$f^{X(1,n,k-1,k),...,X(n,n,k-1,k)}(x_1,...,x_n) = k^n \, n! \prod_{i=1}^{n} (1 - F(x_i))^{k-1} f(x_i)$$

is the joint density function of o OS's based on n iid random variables with distribution function $$G(x) = 1-(1-F(x))^k$$

which is the distribution of the minimum of k random variables distributed according to F. Hence,

$$f^{X(1,n,k-1,k),\dots,X(n,n,k-1,k)}(x_1,\dots,x_n) = f^{Y(1,n,0,1),\dots,Y(n,n,0,1)}(x_1,\dots,x_n)$$

with G being the underlying distribution function of the OS's $Y(j,n,0,1)$, $j = 1,\dots,n$.

One more transformation within the class of g OS's contains a result for records ($m = -1$) (e.g. Nagaraja 1988) :
k–th record values with an underlying distribution function F can be viewed as ordinary record values ($k = 1$) based on the distribution function G of a certain minimum :

$$G(x) = 1-(1-F(x))^k .$$

Remark 3.3.7. Let G be the distribution function of the minimum of $s \in \mathbb{N}$ iid random variables with distribution function F :

$$G(x) = 1-(1-F(x))^s .$$

Then we find

$$f^{X(1,n,s(m+1)-1,sk),\dots,X(n,n,s(m+1)-1,sk)}(x_1,\dots,x_n)$$
$$= f^{Y(1,n,m,k),\dots,Y(n,n,m,k)}(x_1,\dots,x_n)$$

with G being the underlying distribution function of the g OS's $Y(j,n,m,k)$, $j = 1,\dots,n$.

In particular, the transformation of 1–records to k–th records is contained with $m = -1$, $k = 1$, and $s \to k$.

3.4. Concomitants

Proceeding from situations where experiments yield realizations of two quantities $\mathscr{X}$ and $\mathscr{Y}$ and are modelled by bivariate random variables (X,Y), David (1973) introduces 'concomitants' of o OS's in a sample of size n. With respect to one of the marginal random variables, the first, say, order statistics are formed; by this we have an induced order with respect to the second component.

Let (X,Y) , (X_i,Y_i) , $i = 1,..,n$, be bivariate iid random vectors having the joint distribution function $F^{X,Y}$ and let $X_{1,n}$, ... , $X_{n,n}$ be the OS's in the first component. Then the random variable $Y_{[r,n]}$ is assigned to the r–th OS $X_{r,n}$; i.e., if $X_j = X_{r,n}$ for some $j \in \{1,...,n\}$, then $Y_{[r,n]}$ equals Y_j . David (1973) names $Y_{[r,n]}$ the 'concomitant' of $X_{r,n}$; independently, Bhattacharya (1974) introduces the same random variables and terms them 'induced order statistics'.
In the first set–up (see David 1973, 1981), the dependence structure between X und Y is modelled as linear regression. In particular, the case of a bivariate normal distribution of (X,Y) is examined. Moments and product moments are directly obtained depending on the quantities stated in this model. The derivation of the distribution of the rank $R_{r,n}$ of $Y_{[r,n]}$, considering the induced order, is stated in David, O'Connell, Yang (1977). Therein and in David, Galambos (1974) asymptotic properties are shown, too.
Applications of such a model are, e.g., selection, prediction and estimation problems (see David 1981, Galambos 1978, p 267) where inference on the second quantity is intended based on the order statistics. One may also think of certain questions which are vice versa.
In Yang (1977) the special regression model is dropped and a general distribution theory for concomitants of o OS's is established. Densities and joint densities of OS's and concomitants are shown as well as moments and product moments, now appearing as conditional expectations.
For more details on concomitants of o OS's we refer to the review articles of Bhattacharya (1984) and David (1993).
In his dissertation, Houchens (1984) transfers this concept to concomitants of records and presents distribution theory and results on statistical inference.
Concomitants can also be defined in the case of g OS's. As a starting point, the description of g OS's via records from non–identically distributed random variables of Pfeifer (1979, 1982a) (see 1.6.) is chosen noticing Remark 2.8.

Let the family $$\{(X_j^{(r)}, Y_j^{(r)})\}_{1\leq r\leq n\ ,\ j\in\mathbb{N}}$$

of independent, bivariate random vectors with joint distribution function $G_r(x,y)$, $1 \leq r \leq n$, be given; i.e. in each step the underlying distribution may vary and thus there may be different marginal distributions of $Y_1^{(r)}$.

Concerning $X_j^{(r)}$ we consider the construction of records as shown in Section 1.6. with

$$F^{X_1^{(1)}}(x) = 1-(1-F(x))^{k+n-1+M_1} \quad \text{and}$$

$$F^{X_j^{(i)}}(x) = F^{X_1^{(i)}}(x) = 1-(1-F(x))^{k+n-i+M_i}, \quad 2 \le i \le n,\ j \in \mathbb{N}.$$

Determining of records can be viewed as mapping the scheme $(X_j^{(r)})_{1\le r\le n,\ j\in\mathbb{N}}$ into the records. Just the same function, regarding to the first component, is then applied to the scheme $(Y_j^{(r)})_{1\le r\le n,\ j\in\mathbb{N}}$ in order to lead to the bivariate random vectors

$$(X_{\Delta_r}^{(r)}, Y_{[\Delta_r]}^{(r)}),\ 1 \le r \le n;$$

i.e. the concomitants $Y_{[\Delta_r]}^{(r)}$ are assigned to the records $X_{\Delta_r}^{(r)}$ of the first component.

As an example concerning concomitants of g OS's, we derive the joint density of the r–th concomitant $Y_{[\Delta_r]}^{(r)}$ and the r–th record $X_{\Delta_r}^{(r)}$ containing the results in the case of o OS's and ordinary record values.

We observe

$$f^{X_{\Delta_r}^{(r)},\, Y_{[\Delta_r]}^{(r)}}(x,y) = f^{Y_{[\Delta_r]}^{(r)} | X_{\Delta_r}^{(r)}}(y|x)\, f^{X_{\Delta_r}^{(r)}}(x).$$

Because of the independence of the bivariate random vectors together with the identical distribution within any row of the scheme

$$\{(X_j^{(r)}, Y_j^{(r)})\}_{1\le r\le n,\ j\in\mathbb{N}},$$

the conditional density $f^{Y_{[\Delta_r]}^{(r)} | X_{\Delta_r}^{(r)}}$ coincides with the conditional density $f^{Y|X}$ with respect to the underlying distribution. Thus, the results of Yang (1977, p 997) for o OS's and the one of Houchens (1984, p 123) for records are contained.

Using the equivalence $\quad R(\omega) = j \Leftrightarrow X_{\Delta_r}^{(r)}(\omega) = X_j^{(r)}(\omega),$

we obtain

$$P(X^{(r)}_{\Delta_r} \le x\,,\, Y^{(r)}_{[\Delta_r]} \le y)$$

$$= \sum_{j=1}^{\infty} P(X^{(r)}_{\Delta_r} \le x\,,\, Y^{(r)}_{[\Delta_r]} \le y\,,\, R = j)$$

$$= \sum_{j=1}^{\infty} P(X^{(r)}_{j} \le x\,,\, Y^{(r)}_{j} \le y\,,\, R = j)$$

$$= \sum_{j=1}^{\infty} P(X^{(r)}_{1} \le X^{(r-1)}_{\Delta_{r-1}}\,,\, \dots\,,\, X^{(r)}_{j-1} \le X^{(r-1)}_{\Delta_{r-1}}\,,\, X^{(r)}_{j} > X^{(r-1)}_{\Delta_{r-1}}\,,\, Y^{(r)}_{j} \le y\,)$$

$$= \sum_{j=1}^{\infty} \int_{-\infty}^{x} \Big(\int_{-\infty}^{u} P(X^{(r)}_{1} \le v\,,\, \dots\,,\, X^{(r)}_{j-1} \le v \mid X^{(r-1)}_{\Delta_{r-1}} = v)\, dP^{X^{(r-1)}_{\Delta_{r-1}}}(v) \Big)$$

$$\cdot\, P(Y^{(r)}_{1} \le y \mid X^{(r)}_{1} = u)\, f^{X^{(r)}_{1}}(u)\, du$$

$$= \sum_{j=1}^{\infty} \int_{-\infty}^{x} \Big(\int_{-\infty}^{u} (F^{X^{(r)}_{1}}(v))^{j-1}\, dP^{X^{(r-1)}_{\Delta_{r-1}}}(v) \Big)\, P(Y^{(r)}_{1} \le y \mid X^{(r)}_{1} = u)\, f^{X^{(r)}_{1}}(u)\, du$$

$$= \int_{-\infty}^{x} \Big(\int_{-\infty}^{u} (1 - F^{X^{(r)}_{1}}(v))^{-1} f^{X^{(r-1)}_{\Delta_{r-1}}}(v)\, dv \Big)\, P(Y^{(r)}_{1} \le y \mid X^{(r)}_{1} = u)\, f^{X^{(r)}_{1}}(u)\, du\,.$$

Hence, the joint density is given by

$$f^{X^{(r)}_{\Delta_r},\, Y^{(r)}_{[\Delta_r]}}(x,y) = \tfrac{d}{dy}\tfrac{d}{dx} P(X^{(r)}_{\Delta_r} \le x\,,\, Y^{(r)}_{[\Delta_r]} \le y)$$

$$= \tfrac{d}{dy} \Big(\int_{-\infty}^{x} (1 - F^{X^{(r)}_{1}}(v))^{-1} f^{X^{(r-1)}_{\Delta_{r-1}}}(v)\, dv\;\; P(Y^{(r)}_{1} \le y \mid X^{(r)}_{1} = x)\, f^{X^{(r)}_{1}}(x) \Big)$$

$$= \int_{-\infty}^{x} (1 - F^{X^{(r)}_{1}}(v))^{-1} f^{X^{(r-1)}_{\Delta_{r-1}}}(v)\, dv\;\; f^{X^{(r)}_{1}}(x)\, f^{Y^{(r)}_{1} | X^{(r)}_{1}}(y|x)\,.$$

Analogously, using

$$F^{X^{(r)}_{\Delta_r}}(x) = P(X^{(r)}_{\Delta_r} \le x\,,\, Y^{(r)}_{[\Delta_r]} \le \infty) = \sum_{j=1}^{\infty} P(X^{(r)}_{j} \le x\,,\, R = j)$$

$$= \int_{-\infty}^{x} \left(\int_{-\infty}^{u} (1 - F^{X_1^{(r)}}(v))^{-1} f^{X_{\Delta_{r-1}}^{(r-1)}}(v)\, dv \right) f^{X_1^{(r)}}(u)\, du$$

the marginal density of $X_{\Delta_r}^{(r)}$ is obtained :

$$f^{X_{\Delta_r}^{(r)}}(x) = \tfrac{d}{dx} F^{X_{\Delta_r}^{(r)}}(x) = \int_{-\infty}^{x} (1 - F^{X_1^{(r)}}(v))^{-1} f^{X_{\Delta_{r-1}}^{(r-1)}}(v)\, dv \; f^{X_1^{(r)}}(x) .$$

Thus,

$$f^{X_{\Delta_r}^{(r)}, Y_{[\Delta_r]}^{(r)}}(x,y) = f^{X_{\Delta_r}^{(r)}}(x)\, f^{Y_1^{(r)} | X_1^{(r)}}(y | x)$$

and the assertion follows.

Chapter II

Moments of Generalized Order Statistics

One–dimensional marginal density functions and distribution functions have been obtained in the first chapter. In the following two chapters we derive recurrence relations and inequalities for moments of g OS′s and give corresponding characterizations of probability distributions. Hence, we should have at hand some results on the existence of such moments. We state sufficient conditions which are well known in the case of o OS′s and record values (see e.g. David 1981, Sen 1959, Lin 1987). Then we give representations for the difference of moments of successive g OS′s which are used in Chapter III and we derive moments of g OS′s with respect to special distributions. In the second section we cite results on complete function sequences which are a useful tool to obtain characterization results by means of moments.

1. The Existence of Moments of Generalized Order Statistics

Proceeding from the joint density function of the uniform g OS′s $U(j,n,\tilde{m},k)$, $j \in \{1,...,n\}$, the marginal distribution function

$$F^{X(r,n,\tilde{m},k)}(x) = 1 - c_{r-1}\,(1-F(x))^{k+n-r+M_r} \sum_{j=0}^{r-1} \frac{1}{j!\; c_{r-j-1}}\, g_m^j(F(x))$$

of some g OS $X(r,n,\tilde{m},k)$ is derived ($m_1 = ... = m_{r-1} = m$, see I.3.2.6.) where F is an arbitrary distribution function.

Choosing this as a starting point for further examination, we do not need the assumption of absolute continuity imposed on F. E.g., in the representation of an expected value $E\,\Psi(X(r,n,\tilde{m},k))$, Ψ measurable, via the pseudo–inverse F^{-1} of F (see Lemma 1.1.1.) no requirements are needed.

It has previously been noted that in this case the g OS′s $X(r,n,\tilde{m},k)$, $r \in \{1,...,n\}$, may possibly no longer be interpreted as record values, e.g. But conceived as certain random variables possessing the distribution function $F^{X(r,n,\tilde{m},k)}$, strong assumptions on F can be avoided (see I.2.3.).

The pseudo–inverse F^{-1} given by

$$F^{-1}(y) = \inf \{x; F(x) \geq y\}, \quad y \in (0,1)$$

of some distribution function F (Definition I.1.1.2.) is a helpful tool when considering moments of g OS's. Some properties of a pseudo–inverse function are shown in the following lemma (e.g. Witting 1985, p 20).

LEMMA 1.1. Let F be a distribution function and F^{-1} its pseudo–inverse.
Then we have for all $y \in (0,1)$ and for all $x \in \mathbb{R}$:

i) $F(x) \geq y \Leftrightarrow x \geq F^{-1}(y)$,

ii) $F(x-) \leq y \Leftrightarrow x \leq F^{-1}(y+)$,

iii) $F(F^{-1}(y)-) \leq y \leq F(F^{-1}(y))$,

iv) $F^{-1}(F(x)) \leq x \leq F^{-1}(F(x)+)$.

1.1. Sufficient Conditions for the Existence of Moments

Applying the pseudo–inverse, an expression for the expected value of a measurable function of a g OS $X(r,n,\tilde{m},k)$ is obtained; cf. David (1981, p 34), Hwang, Lin (1984a) for o OS's and Grudzień, Szynal (1983) for k–th record values.

LEMMA 1.1.1. Let $X(r,n,\tilde{m},k)$ be a g OS based on the distribution function F, $m_1 = \ldots = m_{r-1} = m$ and let Ψ be a measurable function.
Then we have :

$$E\,\Psi(X(r,n,\tilde{m},k)) = \int_0^1 \Psi(F^{-1}(x))\,\varphi_{r,n}(x)\,dx$$

with $\varphi_{r,n}$ as in I.3.2.1.

Proof Using $f^{U(r,n,\tilde{m},k)}(t) = \varphi_{r,n}(t)\,,\ t \in (0,1)\,,$

we have $\quad E\,\Psi(X(r,n,\tilde{m},k)) = E\,\Psi(F^{-1}(U(r,n,\tilde{m},k))) = \int_0^1 \Psi(F^{-1}(t))\,\varphi_{r,n}(t)\,dt\,.$

Remark 1.1.2. In the sequel, only the existence of the moments $EX^{\alpha}(r,n,\tilde{m},k)$ is required; negative and non–integer valued moments are permitted. Thus, for the moments to be well defined, the following regularity conditions are assumed :

$$\begin{cases} F^{-1}(t) \geq 0 & \text{for all } t \in (0,1)\,,\ \text{if} \quad \alpha > 0\,,\ \alpha \notin \mathbb{N} \\ F^{-1}(t) > 0 & \text{for all } t \in (0,1)\,,\ \text{if} \quad \alpha < 0\,,\ -\alpha \notin \mathbb{N} \\ F^{-1}(t) > 0 & \text{for all } t \in (0,1) \\ \quad \vee\ F^{-1}(t) < 0 & \text{for all } t \in (0,1)\,,\ \text{if} \quad -\alpha \in \mathbb{N} \end{cases} .$$

The existence of some positive moment of a g OS can be ensured globally by the existence of a higher moment (cf. Lin 1987 for records).

Theorem 1.1.3. Let the random variable X be distributed according to F and suppose

$$E|X|^{\beta} < \infty \quad \text{for some } \beta > 0\,.$$

Then for the corresponding g OS's with $m_1 = \ldots = m_{r-1} = m$ we have :

$$E|X(r,n,\tilde{m},k)|^{\alpha} < \infty \quad \text{for all } \alpha \in (0,\beta)\,.$$

Proof The Hölder inequality (see IV.1.), the representation of $\varphi_{r,n}$ (I.3.2.1.) and A.1.2. with $\frac{\beta}{\beta-\alpha} > 0$ together imply

$$E|X(r,n,\tilde{m},k)|^{\alpha} \leq (E|X|^{\beta})^{\alpha/\beta} \left(\int_0^1 (\varphi_{r,n}(t))^{\beta/(\beta-\alpha)}\,dt \right)^{1-\alpha/\beta} < \infty\,.$$

In the above theorem, the existence of $E|X|^{\beta} < \infty$ for some $\beta > 0$ is assumed to guarantee the existence of $E|X(r,n,\tilde{m},k)|^{\alpha} < \infty$ for $0 < \alpha < \beta$.

For o OS's the existence of $E|X|^{\alpha}$ is a sufficient condition; in that case, $\varphi_{r,n}$ is replaced by an upper bound in the representation of Lemma 1.1.1. Thus Lemma I.3.2.12. leads to

COROLLARY 1.1.4. Let F be the distribution function of the random variable X and $X(r,n,\tilde{m},k)$ a corresponding g OS with $m_1 = \dots = m_{r-1} = m$.
If the condition

$$\{r = 1\} \vee \{r > 1 \wedge m > -1\} \vee \{r > 1 \wedge m \leq -1 \wedge k + n - r + M_r > 1\}$$

is satisfied and if $$E|X|^{\alpha} < \infty \quad \text{for some } \alpha \in \mathbb{R},$$

then $$E|X(r,n,\tilde{m},k)|^{\alpha} < \infty .$$

REMARK 1.1.5. The condition in 1.1.4. does not hold for ordinary record values ($m = -1$, $k = 1$); Nagaraja (1978) shows by an example that the existence of EX does not imply the existence of $EX_{L(n)}$. However, Corollary 1.1.4. may be applied to k–th record values with $k > 1$ and $r > 1$.

1.2. The Theorem of Sen for Generalized Order Statistics

In the case of o OS's, the assertion in Corollary 1.1.4. is unsatisfactory since, e.g., the expectations of the o OS's $X_{2,n}, \dots, X_{n-1,n}$ based on a Cauchy distribution exist, whereas the expectation EX itself does not exist. Sen's theorem (Sen 1959) provides a substantial improvement (see also Azlarov, Volodin 1986, p 28/9).

THEOREM 1.2.1. (Sen 1959)
Let X be distributed according to F with a continous density function f.
If $$E|X|^{\delta} < \infty \quad \text{for some } \delta > 0,$$

then for the corresponding o OS's, with $\alpha > 0$, we have :

$$E|X_{r,n}|^{\alpha} < \infty \quad \text{for all } r,n \text{ satisfying } \tfrac{\alpha}{\delta} \leq r \leq n + 1 - \tfrac{\alpha}{\delta}.$$

Let the random variable X be Cauchy distributed; then the moments EX^δ exist for $0 < \delta < 1$ (cf. III.2.3.). Putting $\delta = \frac{\alpha}{\alpha+1}$, $\alpha \in \mathbb{N}$, Sen's theorem yields :

$$E|X_{r,n}|^\alpha < \infty \quad \text{for all } r,n \text{ satisfying } \alpha + 1 \leq r \leq n - \alpha .$$

In the case of g OS's the theorem of Sen can be stated as follows :

THEOREM 1.2.2. Let X be a random variable with distribution function F and continuous density function f satisfying

$$E|X|^\delta < \infty \quad \text{für some } \delta > 0 .$$

If $X(r,n,\tilde{m},k)$ is a g OS based on F with $m_1 = \ldots = m_{r-1} = m$, then, for $\alpha > \delta$, we have :

$$E|X(r,n,\tilde{m},k)|^\alpha < \infty \quad \text{for all } r,n,\tilde{m} \text{ and } k,$$

with

$$r \geq \frac{\alpha}{\delta} \quad \text{and} \quad \begin{cases} k + n - r + M_r \geq \frac{\alpha}{\delta}, & m > -1 \\ k + n - r + M_r > \frac{\alpha}{\delta}, & m = -1 \\ k + n - 1 + M_1 \geq \frac{\alpha}{\delta}, & m < -1 \end{cases} .$$

PROOF In the following representation

$$E|X(r,n,\tilde{m},k)|^\alpha = \int_{-\infty}^{+\infty} |t|^\alpha \varphi_{r,n}(F(t))\, dF(t) = \left(\int_{-\infty}^{0} + \int_{0}^{+\infty}\right) |t|^\alpha \varphi_{r,n}(F(t))\, dF(t)$$

the integrals are considered separately.

Let $c_1(x) = \int_{-\infty}^{x} |t|^\delta dF(t)$; since $c_1(x) < \infty$,

c_1 increases with respect to x and we find $\lim_{x\to-\infty} c_1(x) = 0$,

$\lim_{x\to\infty} c_1(x) = E|X|^\delta < \infty$, and if $x \leq 0$: $|x|^\delta F(x) \leq c_1(x) < \infty$.

We have

$$\int_x^0 |t|^\alpha \varphi_{r,n}(F(t))\, dF(t)$$

$$= \int_x^0 (|t|^\delta F(t))^{\alpha/\delta - 1} F^{1-\alpha/\delta}(t)\, \varphi_{r,n}(F(t))\, |t|^\delta dF(t) = (1), \text{ say.}$$

Now, the function $\Psi(x) = \dfrac{\varphi_{r,n}(x)}{x^{\alpha/\delta-1}}$, continuous in $(0,1)$, is replaced by an upper bound K_1, say. For that purpose, we examine the limits with respect to $x \to 0$ and $x \to 1$.

We have $\lim_{x\to 1} \Psi(x) < \infty$ iff $\lim_{x\to 1} \varphi(x) < \infty$.

Thus, the conditions to be imposed on the parameters can be found in Lemma I.3.2.12. :

$$\{r = 1\} \vee \{r > 1 \wedge m > -1\} \vee \{r > 1 \wedge m \leq -1 \wedge k + n - r + M_r > 1\}.$$

Since $g_m(0) = 0$, we have $\lim_{x\to 0} \Psi(x) < \infty$ for $r \geq \frac{\alpha}{\delta}$.

Hence, the condition

$$\{r \geq \tfrac{\alpha}{\delta} \wedge m > -1\} \vee \{r \geq \tfrac{\alpha}{\delta} \wedge m \leq -1 \wedge k + n - r + M_r > 1\}$$

yields $(1) \leq (c_1(0))^{\alpha/\delta-1} K_1 \int_x^0 |t|^\delta \, dF(t)$.

Thus : $\displaystyle\int_{-\infty}^0 |t|^\alpha \varphi_{r,n}(F(t))\, dF(t) \leq (c_1(0))^{\alpha/\delta-1} K_1 \int_{-\infty}^0 |t|^\delta \, dF(t) < \infty$.

Let $c_2(x) = \int_x^\infty |t|^\delta \, dF(t)$; since $c_2(x) < \infty$,

c_2 decreases with respect to x and we have $\lim_{x\to\infty} c_2(x) = 0$,

$\lim_{x\to-\infty} c_2(x) = E|X|^\delta < \infty$ and if $x \geq 0$: $|x|^\delta (1 - F(x)) \leq c_2(x) < \infty$.

We have

$$\int_0^x |t|^\alpha \varphi_{r,n}(F(t))\, dF(t)$$

$$= \int_0^x (|t|^\delta (1-F(t)))^{\alpha/\delta-1} (1-F(t))^{1-\alpha/\delta} \varphi_{r,n}(F(t))\, |t|^\delta \, dF(t) = (2), \text{ say.}$$

The integrand (up to $|t|^\delta$) is again replaced by an upper bound.

The function $(1-x)^{1-\alpha/\delta} \varphi_{r,n}(x) = \dfrac{c_{r-1}}{(r-1)!} (1-x)^{k+n-r+M_r-\alpha/\delta} g_m^{r-1}(x)$

on $(0,1)$ can be replaced by a finite upper bound K_2, say, (cf. Lemma I.3.2.12.), if

$$\{r = 1\} \vee \{r > 1 \wedge m > -1 \wedge k + n - r + M_r \geq \tfrac{\alpha}{\delta}\}$$
$$\vee \{r > 1 \wedge m = -1 \wedge k + n - r + M_r > \tfrac{\alpha}{\delta}\}$$
$$\vee \{r > 1 \wedge m < -1 \wedge k + n - 1 + M_1 \geq \tfrac{\alpha}{\delta}\}.$$

Thus, $$(2) \leq (c_2(0))^{\alpha/\delta - 1} K_2 \int_0^x |t|^\delta \, dF(t)$$

and hence

$$\int_0^\infty |t|^\alpha \varphi_{r,n}(F(t)) \, dF(t) \leq (c_2(0))^{\alpha/\delta - 1} K_2 \int_0^\infty |t|^\delta \, dF(t) < \infty.$$

Determining the conjunction of the following two conditions,

$$\{r \geq \tfrac{\alpha}{\delta} \wedge m > -1\} \vee \{r \geq \tfrac{\alpha}{\delta} \wedge m \leq -1 \wedge k + n - r + M_r > 1\}$$

and $$\{r = 1\} \vee \{r > 1 \wedge m > -1 \wedge k + n - r + M_r \geq \tfrac{\alpha}{\delta}\}$$
$$\vee \{r > 1 \wedge m = -1 \wedge k + n - r + M_r > \tfrac{\alpha}{\delta}\}$$
$$\vee \{r > 1 \wedge m < -1 \wedge k + n - 1 + M_1 \geq \tfrac{\alpha}{\delta}\},$$

the assertion of the theorem follows.

Again, ordinary record values ($m = -1$, $k = 1$) do not fulfil the condition, since $k + n - r + M_r = k = 1 \leq \frac{\alpha}{\delta}$.

Sen's theorem (Sen 1959) is contained in Theorem 1.2.2. putting $m = 0$ and $k = 1$.

Example to Theorem 1.2.2. with $m = -1$.

The expectation of a Pareto distributed random variable X having distribution function

$$F(x) = \begin{cases} 0 & , x \leq 1 \\ 1 - 1/x & , x > 1 \end{cases}$$

does not exist.

Since $E|X|^\delta < \infty$, if $\delta < 1$, the expectation $E|X(r,n,-1,k)|$ exists, if $r \geq \frac{1}{\delta}$ and $k > \frac{1}{\delta}$, however; hence, $E|X(r,n,-1,k)|$ exists, if $r,k \geq 2$.

(Moreover, $E|X(1,n,-1,k)|$ exists ($= \frac{k}{k-1}$ see 1.4.2.).)

REMARK 1.2.3. Upper bounds for the functions $\varphi_{r,n}$ in I.3.2.1. (cf. I.3.2.12.) lead to inequalities between absolute moments of g OS's and corresponding moments with respect to the underlying distribution (see IV.1.).

Example in the case $m_1 = \ldots = m_{n-1} = -1$ and $r,k > 1$:

The function $\qquad \varphi_{r,n}(x) = \frac{k^r}{(r-1)!}(1-x)^{k-1}(\log \frac{1}{1-x})^{r-1}$, $x \in (0,1)$,

attains its maximum at $\qquad x^* = 1 - \exp(-\frac{r-1}{k-1})$

with $\qquad \varphi_{r,n}(x^*) = \frac{k^r}{(r-1)!} e^{1-r} (\frac{r-1}{k-1})^{r-1}$.

Thus we have (subject to the existence of $E|X|^{\alpha}$) :

$$E|X(r,n,-1,k)|^{\alpha} = \int_0^1 |F^{-1}(t)|^{\alpha} \varphi_{r,n}(t)\, dt \leq \frac{k^r}{(r-1)!} e^{1-r} (\tfrac{r-1}{k-1})^{r-1} E|X|^{\alpha} .$$

1.3. Differences of Moments

Under different conditions imposed on the underlying distribution function F (absolute continuity of F^{-1} or continuity of F , see the introductory remarks in the first section), representations for the difference of moments of successive g OS's are now given.

THEOREM 1.3.1. Let F be a distribution function possessing an absolutely continuous pseudo-inverse F^{-1}, $r \geq 2$, and let the regularity condition

$$\begin{cases} F^{-1}(t) \geq 0 & \text{for all } t \in (0,1) \text{ , if } \quad \alpha > 1 \text{ , } \alpha \notin \mathbb{N} \\ F^{-1}(t) > 0 & \text{for all } t \in (0,1) \text{ , if } \quad \alpha < 1 \text{ , } -\alpha \notin \mathbb{N} \\ F^{-1}(t) > 0 & \text{for all } t \in (0,1) \\ \quad \vee\ F^{-1}(t) < 0 & \text{for all } t \in (0,1) \text{ , if } \quad -\alpha \in \mathbb{N} \end{cases}$$

be fulfilled with respect to $0 \neq \alpha \in \mathbb{R}$ and F . Moreover, let $m_1 = \ldots = m_{r-1} = m$.
If the moments $EX^{\alpha}(r,n,\tilde{m},k)$ and $EX^{\alpha}(r-1,n,\tilde{m},k)$ of g OS's based on F exist, then

$EX^{\alpha}(r,n,\tilde{m},k) - EX^{\alpha}(r-1,n,\tilde{m},k)$

$$= \alpha \frac{c_{r-2}}{(r-1)!} \int_0^1 (F^{-1}(t))^{\alpha-1} (F^{-1})'(t)\,(1-t)^{k+n-r+M_r}\, g_m^{r-1}(t)\, dt\,.$$

Proof The representation is proved via integration by parts. For that purpose, two auxiliary properties are shown :

1. $$\lim_{t\to 1-} (F^{-1}(t))^{\alpha}(1-t)^{k+n-r+M_r} g_m^{r-1}(t) = 0 \quad \text{and}$$

2. $$\lim_{t\to 0+} (F^{-1}(t))^{\alpha}(1-t)^{k+n-r+M_r} g_m^{r-1}(t) = 0\,.$$

ad 1.: The function $\varphi_{r,n}$ (see I.3.2.1.) is a density function on (0,1); thus :

$\lim_{t\to 1-} (1-t)^{k+n-r+M_r} g_m^{r-1}(t) = 0$, since

$$0 = \lim_{t\to 1-} \int_t^1 (1-x)^{k+n-r+M_r-1} g_m^{r-1}(x)\, dx$$

$$\geq \lim_{t\to 1-} g_m^{r-1}(t) \int_t^1 (1-x)^{k+n-r+M_r-1} dx \geq 0\,.$$

Moreover, the expected value $EX^{\alpha}(r,n,\tilde{m},k)$ exists; hence

$$0 = \lim_{t\to 1-} \int_t^1 (F^{-1}(x))^{\alpha}(1-x)^{k+n-r+M_r-1} g_m^{r-1}(x)\, dx = (1)\,,\ \text{say.}$$

If $(F^{-1}(t))^{\alpha}$ is bounded, $t \to 1-$, so 1. is obvious.

Let $\lim_{t\to 1-} (F^{-1}(t))^{\alpha} = \pm\infty$.

1.1. $\alpha > 0$, $F^{-1}(t) \to \infty$, $t \to 1-$:

$F^{-1}(t) > 0$ in a neighbourhood on the left of 1 with $(F^{-1}(t))^{\alpha}$ monotonically increasing.

Thus $(F^{-1}(t))^{\alpha} g_m^{r-1}(t)$ is monotonically increasing, too, and noticing

$$(1) \geq \lim_{t\to 1-} (F^{-1}(t))^{\alpha} g_m^{r-1}(t) \int_t^1 (1-x)^{k+n-r+M_r-1} dx$$

$$= \lim_{t\to 1-} (F^{-1}(t))^{\alpha} g_m^{r-1}(t) \frac{1}{k+n-r+M_r} (1-t)^{k+n-r+M_r} \geq 0\,,$$

the assertion follows.

1.2. $\alpha < 0$, $F^{-1}(t) \to 0$, $t \to 1-$ (i.e. $F^{-1}(x) < 0$ for all $x \in (0,1)$ and $-\alpha \in \mathbb{N}$) :

If α is even, $(F^{-1}(x))^{\alpha} > 0$ and monotonically increasing; thus the argumentation is that of 1.1.

If α is odd, $(F^{-1}(x))^{\alpha} < 0$ and monotonically decreasing. Then $(F^{-1}(x))^{\alpha} \cdot g_m^{r-1}(x)$ is monotonically decreasing, too, and noticing

$$(1) \leq \lim_{t \to 1-} (F^{-1}(t))^{\alpha} g_m^{r-1}(t) \int_t^1 (1-x)^{k+n-r+M_r-1} dx \leq 0,$$

the assertion follows.

ad 2.: The expectation $EX^{\alpha}(r-1,n,\tilde{m},k)$ exists; hence

$$0 = \lim_{t \to 0+} \int_0^t (F^{-1}(x))^{\alpha} (1-x)^{k+n-r+M_r+m} g_m^{r-2}(x)\, dx = (2), \text{ say.}$$

Moreover, $\lim_{t \to 0+} (1-t)^{k+n-r+M_r} g_m^{r-1}(t) = 0$.

If $(F^{-1}(t))^{\alpha}$ is bounded, $t \to 0+$, so 2. is obvious.

Let $\lim_{t \to 0+} (F^{-1}(t))^{\alpha} = \pm \infty$.

2.1. $\alpha < 0$, $F^{-1}(t) \to 0$, $t \to 0+$:

$F^{-1}(t) > 0$ for all $t \in (0,1)$, $(F^{-1}(t))^{\alpha}$ monotonically decreasing, hence $(F^{-1}(t))^{\alpha} (1-t)^{k+n-r+M_r}$ monotonically decreasing.

Noticing that

$$(2) \geq \lim_{t \to 0+} (F^{-1}(t))^{\alpha} (1-t)^{k+n-r+M_r} \int_0^t (1-x)^m g_m^{r-2}(x)\, dx$$
$$= \lim_{t \to 0+} \frac{1}{r-1} (F^{-1}(t))^{\alpha} (1-t)^{k+n-r+M_r} g_m^{r-1}(t) \geq 0,$$

the assertion follows.

2.2. $\alpha > 0$, $F^{-1}(t) \to -\infty$, $t \to 0+$ (i.e. $\alpha \in \mathbb{N}$) :

$F^{-1}(t) < 0$ in a neighbourhood on the right of 0.

If α is even, $(F^{-1}(t))^{\alpha} > 0$, $(F^{-1}(t))^{\alpha}$ and hence, $(F^{-1}(t))^{\alpha} (1-t)^{k+n-r+M_r}$ monotonically decreasing; the assertion follows as in 2.1.

If α is odd, $(F^{-1}(t))^\alpha < 0$ and $(F^{-1}(t))^\alpha$ monotonically increasing; since $(F^{-1}(t))^\alpha (1-t)^{k+n-r+M_r}$ is monotonically increasing, too, and

$$(2) \leq \lim_{t \to 0+} (F^{-1}(t))^\alpha (1-t)^{k+n-r+M_r} \int_0^t (1-x)^m g_m^{r-2}(x)\, dx \leq 0,$$

2. is proved.

The assertion of the theorem is now proved via integration by parts using Lemma 1.1.1. and Corollary I.3.2.8. :

$$\begin{aligned}
&EX^\alpha(r,n,\tilde{m},k) - EX^\alpha(r-1,n,\tilde{m},k) \\
&\quad = \int_0^1 (F^{-1}(t))^\alpha (\varphi_{r,n}(t) - \varphi_{r-1,n}(t))\, dt \\
&\quad = \int_0^1 (F^{-1}(t))^\alpha\, d(\Phi_{r,n}(t) - \Phi_{r-1,n}(t)) \\
&\quad = -\frac{c_{r-2}}{(r-1)!} (F^{-1}(t))^\alpha (1-t)^{k+n-r+M_r} g_m^{r-1}(t) \Big|_0^1 \\
&\qquad + \frac{c_{r-2}}{(r-1)!} \int_0^1 (1-t)^{k+n-r+M_r} g_m^{r-1}(t)\, d((F^{-1}(t))^\alpha) \\
&\quad = \alpha \frac{c_{r-2}}{(r-1)!} \int_0^1 (F^{-1}(t))^{\alpha-1} (F^{-1})'(t)\, (1-t)^{k+n-r+M_r} g_m^{r-1}(t)\, dt .
\end{aligned}$$

In the case of o OS's this representation is given by Lin (1988b) to obtain recurrence relations for moments (see III.2.). Under modified conditions and without use of the pseudo–inverse such an expression is noted in David, Shu (1978) and Khan, Yaqub, Parvez (1983) for $\alpha \in \mathbb{N}$. This version is stated in Theorem 1.3.4. for g OS's; only the continuity of F is required. An expression for the difference of distribution functions of successive g OS's is known from Corollary I.3.2.10. Applying a representation of an expected value as a difference of integrals, the desired result is obtained.

If $E|X| < \infty$ is satisfied for some random variable X, the expectation may be derived using the formula

$$EX = \int_0^\infty (1 - F(x))\, dx - \int_{-\infty}^0 F(x)\, dx$$

(e.g. Rohatgi 1976).

Summarizing, the following lemma holds for arbitrary moments (see e.g. Cacoullos 1989 for $\alpha \in \mathbb{N}$).

Lemma 1.3.2. Let the random variable X be distributed according to the continuous distribution function F with

$$E|X^{\alpha}| < \infty \text{ for some } \alpha \in \mathbb{R}, \alpha \neq 0 .$$

Then, under the regularity conditions of Theorem 1.3.1., we find the following representation of the moment EX^{α} :

$$\begin{cases} \alpha \int_0^{\infty} x^{\alpha-1} (1 - F(x))\, dx - \alpha \int_{-\infty}^{0} x^{\alpha-1} F(x)\, dx , & \alpha \in \mathbb{N} , \text{ F arbitrary} \\ \alpha \int_0^{\infty} x^{\alpha-1} (1 - F(x))\, dx , & \alpha > 0 , \alpha \notin \mathbb{N} \\ -\alpha \int_0^{\infty} x^{\alpha-1} F(x)\, dx , & \alpha < 0 , -\alpha \notin \mathbb{N} \\ -\alpha \int_0^{\infty} x^{\alpha-1} F(x)\, dx , & -\alpha \in \mathbb{N} , F^{-1}(t) > 0 \\ \alpha \int_{-\infty}^{0} x^{\alpha-1} (1 - F(x))\, dx , & -\alpha \in \mathbb{N} , F^{-1}(t) < 0 \end{cases}$$

Lemma 1.3.2. implies the following representation for the difference of two moments of order α.

Corollary 1.3.3. Let X and Y be distributed according to continuous distribution functions F and G, respectively. Let the supports of F and G coincide and $E|X^{\alpha}| < \infty$, $E|Y^{\alpha}| < \infty$ for some $\alpha \in \mathbb{R}$, $\alpha \neq 0$.

If the regularity conditions of 1.3.2. hold for F and G, then

$$EX^{\alpha} - EY^{\alpha} = \alpha \int_{-\infty}^{\infty} x^{\alpha-1} (G(x) - F(x))\, dx .$$

Theorem 1.3.4. Let the distribution function F be continuous, $r \geq 2$, and let the regularity conditions of Theorem 1.3.1. be fulfilled with respect to $\alpha \in \mathbb{R}$, $\alpha \neq 0$ and F .

Moreover, let the moments $EX^{\alpha}(r,n,\tilde{m},k)$ and $EX^{\alpha}(r-1,n,\tilde{m},k)$ of g OS's based on F exist and $m_1 = \dots = m_{r-1} = m$.

Then, with $a = F^{-1}(0)$, $b = F^{-1}(1)$, we find :

$$EX^{\alpha}(r,n,\tilde{m},k) - EX^{\alpha}(r-1,n,\tilde{m},k) = \alpha \frac{c_{r-2}}{(r-1)!} \int_a^b x^{\alpha-1} (1-F(x))^{k+n-r+M_r} g_m^{r-1}(F(x))\, dx\,.$$

If in Theorem 1.3.1. F is further assumed to be absolutely continuous (as required in the form of the joint density of g OS's in I.2.4.), Theorem 1.3.4. follows directly via the transformation $t = F(x)$.

Proceeding from the marginal distribution functions of the g OS's $X(r,n,\tilde{m},k)$ and $X(r-1,n,\tilde{m},k)$, we are interested in rather weak conditions imposed on F and F^{-1}, respectively. Thus, in the above theorem only continuity of F is assumed.

1.4. Moments for Specific Distributions

To show some examples, explicit expressions and identities for moments of g OS's are derived in the case of power function, Pareto and Weibull distributions.

1.4.1. Power function distributions

distribution function	$F(x) = c\,x^a,\ x \in (0,c^{-1/a}),\ a,c > 0,$
pseudo–inverse	$F^{-1}(t) = (\frac{t}{c})^{1/a},\ t \in (0,1)\,.$

For $\frac{\alpha}{a} \in \mathbb{N}_0$ we find the following representation :

$$EX^{\alpha}(r,n,m,k) = c_{r-1}(k)\, c^{-\alpha/a} \sum_{j=0}^{\alpha/a} (-1)^j \binom{\alpha/a}{j} \frac{1}{c_{r-1}(k+j)}\,.$$

Restricting to $m \neq -1$, $\frac{\alpha}{a} + 1 > 0$, we have :

$$EX^{\alpha}(r,n,m,k) = \frac{c_{r-1}}{(r-1)!} (m+1)^{1-r}\, c^{-\alpha/a} \sum_{j=0}^{r-1} (-1)^j \binom{r-1}{j} \frac{\Gamma(\alpha/a+1)\ \Gamma(k+(n-r+j)(m+1))}{\Gamma(\alpha/a+1+k+(n-r+j)(m+1))}\,.$$

In particular for $a = 1$, i.e. uniform distribution, we find :

$$EX(r,n,m,k) - EX(r-1,n,m,k) = \frac{1}{c}\, \frac{c_{r-2}(k)}{c_{r-1}(k+1)}\,.$$

Examples :

$m = 0, k = 1$:

$$EX^{\alpha}(r,n,0,1) = r\binom{n}{r} c^{-\alpha/a} \sum_{j=0}^{r-1} (-1)^j \binom{r-1}{j} \frac{\Gamma(\alpha/a+1)\ \Gamma(n-r+j+1)}{\Gamma(\alpha/a+n-r+j+2)} .$$

$$= c^{-\alpha/a} \frac{n!\,\Gamma(\alpha/a+r)}{(r-1)!\ \Gamma(\alpha/a+n+1)} \quad \text{(Malik 1967)} .$$

$m = -1$, $k \in \mathbb{N}$, $\alpha/a \in \mathbb{N}$:

$$EX^{\alpha}(r,n,-1,k) = k^r c^{-\alpha/a} \sum_{j=0}^{\alpha/a} (-1)^j \binom{\alpha/a}{j} (k+j)^{-r}$$

($\alpha = 1,2$, $k = 1$ in Ahsanullah 1989) .

In the case of a uniform distribution ($a = 1$) we have :

$$EX^{\alpha}(r,n,-1,k) = k^r c^{-\alpha} \sum_{j=0}^{\alpha} (-1)^j \binom{\alpha}{j} (k+j)^{-r} ,$$

hence

$$EX(r,n,-1,k) = \tfrac{1}{c} k^r (k^{-r} - (k+1)^{-r})$$

(see Grudzień, Szynal 1983, $k = 1$ in Nagaraja 1978) .

1.4.2. Pareto distributions

distribution function $F(x) = 1 - c/x^a$, $x \in (c^{1/a},\infty)$, $a,c > 0$,

pseudo–inverse $F^{-1}(t) = (\tfrac{1}{c}(1-t))^{-1/a}$, $t \in (0,1)$.

If α and a satisfy the conditions

$$k+(n-r)(m+1) > \alpha/a \quad \text{and} \quad k+(n-1)(m+1) > \alpha/a ,$$

then we have the following representation :

$$EX^{\alpha}(r,n,m,k) = c^{\alpha/a} \frac{c_{r-1}(k)}{c_{r-1}(k-\alpha/a)} .$$

Examples :

$m = 0, k = 1$:

$$EX^{\alpha}(r,n,0,1) = c^{\alpha/a} \frac{n!}{(n-r)!} \frac{\Gamma(n-r+1-\alpha/a)}{\Gamma(n+1-\alpha/a)} \quad \text{(Malik 1966)} .$$

$m = -1$, $k \in \mathbb{N}$:

$$EX^{\alpha}(r,n,-1,k) = c^{\alpha/a} \frac{k^r}{(k-\alpha/a)^r} \quad \text{and hence}$$

$$EX(r,n,-1,k) = c^{1/a} \frac{k^r}{(k-1/a)^r}$$

(see Grudzień, Szynal 1983, k = 1 in Nagaraja 1978, Ahsanullah, Houchens 1989) .

1.4.3. Weibull distributions

distribution function $F(x) = 1 - \exp\{-c\, x^a\}$, $x \in (0,\infty)$, $a,c > 0$,

pseudo–inverse $F^{-1}(t) = (\frac{1}{c} \log \frac{1}{1-t})^{1/a}$, $t \in (0,1)$.

For $\alpha/a + 1 > 0$ we have :

$EX^{\alpha}(r,n,m,k)$

$$= \begin{cases} \frac{c_{r-1}}{(r-1)!} (m+1)^{1-r}\, c^{-\alpha/a}\, \Gamma(\alpha/a+1) \sum_{j=0}^{r-1} (-1)^j \binom{r-1}{j} (k+(n-r+j)(m+1))^{-\alpha/a-1} & , m \neq -1 \\ \frac{c_{r-1}}{(r-1)!}\, c^{-\alpha/a}\, \Gamma(\alpha/a+r)\, k^{-\alpha/a-r} & , m = -1 \end{cases} .$$

In the special case $a = 1$, i.e. exponential distribution, we find $(EX(0,n,m,k) = 0)$:

$$EX(r,n,m,k) - EX(r-1,n,m,k) = \frac{1}{c} \frac{1}{k+(n-r)(m+1)} .$$

Examples :

$m = 0$, $k = 1$:

$$EX^{\alpha}(r,n,0,1) = r\binom{n}{r} c^{-\alpha/a}\, \Gamma(\alpha/a+1) \sum_{j=0}^{r-1} (-1)^j \binom{r-1}{j} (n-r+j+1)^{-\alpha/a-1}$$

(Lieblein 1955).

$m = -1$, $k \in \mathbb{N}$:

$$EX^{\alpha}(r,n,-1,k) = \frac{1}{(r-1)!} c^{-\alpha/a}\, \Gamma(\alpha/a+r)\, k^{-\alpha/a} \quad \text{and hence}$$

$$EX(r,n,-1,k) = \frac{1}{(r-1)!} c^{-1/a}\, \Gamma(1/a+r)\, k^{-1/a}$$

(see Grudzień, Szynal 1983, k = 1 in Nagaraja 1978) .

Thus, for an exponential distribution $(a = 1)$ we find :

$$EX(r,n,-1,k) = \frac{r}{c\,k} \quad (k = 1 \text{ in Ahsanullah 1987}) .$$

2. Characterization of Distributions by Sequences of Moments

Hoeffding (1953) shows that the expected values of order statistics

$$(EX_{r,n})_{1\leq r\leq n,\ n\in\mathbb{N}}$$

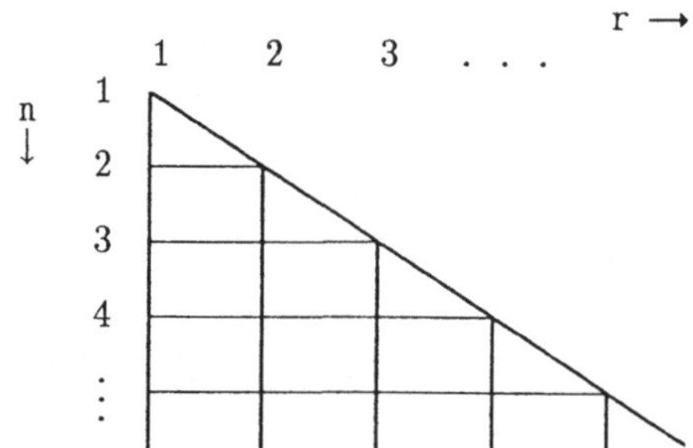

in this triangular array characterize the underlying distribution function, if the first absolute moment exists.

The assertion remains valid, if only the sequence of minima $(EX_{1,n})_{n\in\mathbb{N}}$ or if the sequence of maxima $(EX_{n,n})_{n\in\mathbb{N}}$ is known (Chan 1967, Konheim 1971). Pollak (1973) assumes knowledge of some subsequence $(EX_{r(n),n})_{n\in\mathbb{N}}$, choosing for each n some r(n) with $1 \leq r(n) \leq n$.

However, by this the original assumption of Hoeffding is not really weakened, since the whole triangular array of expectations of the order statistics can be reconstructed by the cited sequences of moments via the following well–known recurrence relation which is valid for arbitrary distributions :

$$(n-r)\, EX^{\alpha}_{r,n} + r\, EX^{\alpha}_{r+1,n} = n\, EX^{\alpha}_{r,n-1}, \quad 1 \leq r \leq n-1 ;$$

(see Section III.1.); this fact is pointed out by Mallows (1973) and Kadane (1974).

In the case of record values, Kirmani, Beg (1984) show that the sequence of expectations $(EX_{L(n)})_{n\geq n_0}$ characterizes the underlying continuous distribution function, if the existence of some moment of order $p > 1$ is assumed.

2.1. Complete Function Sequences

The connection between the introductory remarks and the completeness of certain function sequences is obvious by considering the definition (cf. Hwang, Lin 1984a, Lin 1989a):

DEFINITION 2.1.1. (cf. Hwang, Lin 1984a)
Let $L_\delta(A)$ be the space of δ–integrable functions on a measurable set $A \subset \mathbb{R}$. A sequence $(f_n)_{n\in\mathbb{N}}$ of functions in $L_\delta(A)$ is called complete on $L_\delta(A)$, if for all functions $g \in L_\delta(A)$ the condition

$$\int_A g(x)\, f_n(x)\, dx = 0 \quad \text{for all } n \in \mathbb{N}$$

implies : $$g(x) = 0 \quad \text{a.e. on } A\,.$$

Notations : $L_\delta(A) = L_\delta(a,b)$, if $A = (a,b)$; $L(a,b) = L_1(a,b)$.

The connection to characterizations of probability distributions is indicated by the following :

Let the random variables X and Y be distributed according to F and G, respectively. Thus, if the expectations $EX_{r,n}$ and $EY_{r,n}$ of OS's coincide for some sequence of indices,

$$\int_0^1 \left(F^{-1}(t) - G^{-1}(t)\right) t^{r-1} (1-t)^{n-r}\, dt = 0$$

yields the equality of F and G via a complete function sequence.

In the following, some results for complete function sequences are listed.

The Müntz–Szász lemma permits the choice of a real subsequence of Hoeffding's triangular scheme; this result dates back to Müntz (1914), Szász (1916) and is cited, e.g., in Boas (1954), Hwang, Lin (1984a) and Lin (1989a).

LEMMA 2.1.2. (Müntz, Szász) Let $(n_j)_{j\in\mathbb{N}} \subset \mathbb{N}$ with $n_1 < n_2 < \dots$.
Then the sequence $(x^{n_j})_{j\in\mathbb{N}}$ of polynomials is complete on $L(0,1)$, iff $\sum_{j=1}^{\infty} n_j^{-1} = \infty$.

A generalization of this lemma is given in

Lemma 2.1.3. (Hwang 1983)
Let f be an absolutely continuous function on a bounded interval $[a,b]$ with $f(a)\,f(b) \geq 0$ and $|f'(x)| \geq k > 0$ a.e. on $[a,b]$.

Moreover, let $(n_j)_{j\in\mathbb{N}} \subset \mathbb{N}$ be a subsequence of $\mathbb{N}$ with $n_1 < n_2 < \ldots$ and $\sum_{j=1}^{\infty} n_j^{-1} = \infty$.

Then the function sequence $(f^{n_j}(x))_{j\in\mathbb{N}}$ is complete on $L(a,b)$, iff f strictly increases on $[a,b]$.

Remark 2.1.4. Lemma 2.1.3. remains valid, if the assumption

$$|f'(x)| \geq k > 0 \quad \text{a.e. on } [a,b]$$

is replaced by

$$f'(x) \neq 0 \quad \text{a.e. on } (a,b)$$

(cf. Hwang, Lin 1984a,b).

In order to modify the condition imposed on the function f, a theorem of Zaretzki (cf. Natanson 1961) can be applied.

Theorem 2.1.5. (Zaretzki)
Let f be a continuous and strictly increasing function on a bounded interval $[a,b]$. Then f^{-1} is absolutely continuous on $[f(a),f(b)]$, iff $f'(x) \neq 0$ a.e. on (a,b).

The following lemma shows some results on the completeness of sequences

$$(\, x^{r-1}\,(1-x)^{n-r}\,)_{(r,n)\in I_j}\,.$$

Lemma 2.1.6. For any set I_j of pairs of indices (r,n), $1 \leq r \leq n$, the sequence of polynomials $(\, x^{r-1}\,(1-x)^{n-r}\,)_{(r,n)\in I_j}$ is complete on $L(0,1)$:

$I_1 = \{ (r,n) \mid$ let $\mu \in \mathbb{N}$ fixed; for each $n \geq \mu$ choose some $r = r_n$, $1 \leq r_n \leq n$, with $r_\mu \leq r_n \leq r_\mu + n - \mu \}$ (Huang 1975),

$I_2 = \{ (r,n) \mid$ for each $n \geq 2$ choose some $r = r_n$ with $1 \leq r_n \leq n \}$ (Huang, Hwang 1975, see Hwang, Lin 1984a, p 187),

$I_3 = \{ (r,n) \mid$ each $n \in (n_j)_{j\in\mathbb{N}}$ with $n_j \to \infty$, $j \to \infty$, is combined with all $r = r_j$, $r_j \in \{1,\ldots,n_j\} \}$ (Hwang 1978),

$I_4 = \{ (r,n) \mid$ for given sequences $(n_j)_{j\in\mathbb{N}}, (\mu_j)_{j\in\mathbb{N}} \subset \mathbb{N}$ satisfying $\mu_{j+1} > n_j \geq \mu_j > 1$, $j \in \mathbb{N}$ and $\sum_{j=1}^{\infty} \sum_{\nu=\mu_j}^{n_j} \frac{1}{\nu-1} = \infty$, each $n \in (n_j)_{j\in\mathbb{N}}$ is combined with all $r = r_j$, $r_j \in \{\mu_j,\ldots,n_j\} \}$ (Hwang, Lin 1984a).

As in Lemma 2.1.3., a result on the completeness of

$$(f^{r-1}(x)\,(1 - f(x))^{n-r})_{(r,n)}$$

with respect to suitable functions f is stated.

Lemma 2.1.7. Let f be an absolutely continuous function on $[a,b]$ and $f'(x) \neq 0$ a.e. on (a,b). Moreover, let I be a set of indices (r,n), such that the sequence of polynomials

$$(x^{r-1}\,(1 - x)^{n-r})_{(r,n)\in I}$$

is complete on $L(f(a),f(b))$.

Then if f strictly increases on (a,b),

$$(f^{r-1}(x)\,(1 - f(x))^{n-r})_{(r,n)\in I}$$

is complete on $L(a,b)$.

Proof Let $g \in L(a,b)$ be arbitrary

with $$\int_a^b f^{r-1}(x)\,(1 - f(x))^{n-r}\,g(x)\,dx = 0 \quad \text{for all } (r,n) \in I.$$

Since $f'(x) \neq 0$ a.e. on (a,b), f^{-1} is absolutely continuous on $[f(a),f(b)]$ (Zaretzki's theorem).

Putting $x = f^{-1}(y)$, we have

$$\int_a^b f^{r-1}(x)\,(1-f(x))^{n-r}\,g(x)\,dx = \int_{f(a)}^{f(b)} y^{r-1}\,(1-y)^{n-r}\,g(f^{-1}(y))\,(f^{-1})'(y)\,dy = (*)\,.$$

Moreover, $g(f^{-1}(y))\,(f^{-1})'(y) \in L(f(a),f(b))$, because of

$$\left|\int_{f(a)}^{f(b)} g(f^{-1}(y))\,(f^{-1})'(y)\,dy\right| = \left|\int_a^b g(x)\,dx\right| < \infty \quad (\,g \in L(a,b)\,).$$

Now, $(*) = 0$ for all $(r,n) \in I$.

Hence, $g(f^{-1}(y))\,(f^{-1})'(y) = 0$ a.e. on $(f(a),f(b))$ because of the completeness of

$$(\,x^{r-1}\,(1-x)^{n-r}\,)_{(r,n)\in I} \text{ on } L(f(a),f(b))\,.$$

Strict monotonicity and absolute continuity of f on $[a,b]$ together with Zaretzki's theorem imply $(f^{-1})'(x) \neq 0$ a.e. on $(f(a),f(b))$.

The identity $g(f^{-1}(y)) = 0$ a.e. on $(f(a),f(b))$ implies $g(x) = 0$ a.e. on (a,b); thus the assertion is proved with $g \in L(a,b)$ being arbitrary.

To prepare for results on records ($m = -1$) (cf. Section 2.2.), some other complete function sequences are shown.

LEMMA 2.1.8. (Lin 1987)

For each $p > 1$ the function sequence $(\,(\log \frac{1}{1-x})^n\,)_{n\in\mathbb{N}_0}$ is complete on $L_p(0,1)$.

If this lemma is to be applied, the validity of a certain relationship must hold for all $n \in \mathbb{N}_0$. In a modification, $n = 0,1,\ldots,n_0-1$, $n_0 \in \mathbb{N}$ fixed, may be dropped corresponding to the completeness on $L(0,\delta)$, $0 < \delta < 1$.

LEMMA 2.1.9. (Lin 1986)

For each fixed $n_0 \in \mathbb{N}_0$ and for each fixed $a \geq 0$ the function sequence $(\,x^n\,e^{-x}\,)_{n\geq n_0}$ is complete on $L(a,\infty)$.

A restriction on a subsequence of $\mathbb{N}$ just as in the Müntz–Szász lemma is not possible in general; an example is given in Lin (1986).
Lemma 2.1.9. and Zaretzki's theorem together imply

LEMMA 2.1.10. (Lin 1986)
Let $n_0 \in \mathbb{N}_0$ fixed, $-\infty \leq a < b \leq \infty$ and let f be an absolutely continuous function on $[a,b]$ with $f(x) \geq 0$ and $f'(x) \neq 0$ a.e. on (a,b).
Then the function sequence $(\, f^n(x)\, e^{-f(x)}\,)_{n \geq n_0}$ is complete on $L(a,b)$, iff f is strictly increasing on (a,b).

COROLLARY 2.1.11. For each fixed $n_0 \in \mathbb{N}_0$ the function sequence

$$\left(\, (\log \tfrac{1}{1-x})^n\, (1-x)\, \right)_{n \geq n_0}$$

is complete on $L(0,\delta)$, $0 < \delta < 1$.

2.2. Characterizing Sequences of Expectations

Characterizing sequences of moments of o OS's and record values have previously been mentioned. Here, we state a version for g OS's followed by a result on moment differences.

LEMMA 2.2.1. (e.g. Hwang, Lin 1984a)
Let F and G be distribution functions.
Then $F^{-1}(t) = G^{-1}(t)$ a.e. on $(0,1)$ implies $F(x) = G(x)$ for all $x \in \mathbb{R}$.

REMARK 2.2.2. In the case of continuous distribution functions, Lemma 2.2.1. can be extended to powers of the pseudo–inverses (cf. Lin 1987).
There exists $\nu \in \mathbb{N}$ satisfying $(F^{-1}(t))^\nu = (G^{-1}(t))^\nu$ a.e. on $(0,1)$, iff $F(x) = G(x)$ for all $x \in \mathbb{R}$.

If $\nu \in \mathbb{N}$ is even, the assertion holds due to the fact that the pseudo–inverse of a continuous distribution function increases strictly and is continuous on the left on $(0,1)$.

THEOREM 2.2.3. Let F be a distribution function with an absolutely continuous pseudo–inverse F^{-1}. Let I be a set of pairs $(r,n) \in \mathbb{N}^2$ such that the function sequence

$$\begin{cases} \left((t^{m+1})^{n-r}\,(1-t^{m+1})^{r-1}\right)_{(r,n)\in I} & \text{is complete on } L(0,1)\,,\ \text{if } m>-1 \\ \left((\log\frac{1}{1-t})^{r-1}\right)_{(r,n)\in I} & \text{is complete on } L_\delta(0,1)\,,\ \delta>1\,,\ \text{if } m=-1 \end{cases}$$

If for some fixed parameters k and m

$$\begin{cases} F^{-1}(t)\,(1-t)^{k-1} \in L(0,1)\,, & \text{if } m>-1 \\ F^{-1}(t)\,(1-t)^{k-1} \in L_\delta(0,1)\,, & \text{if } m=-1 \end{cases},$$

then the sequence of expectations

$$(\ EX(r,n,m,k)\ )_{(r,n)\in I}$$

characterizes the distribution function F, if the existence of the expectations is ensured.

PROOF Let X(r,n,m,k) and Y(r,n,m,k) be g OS's based on F and G, respectively.

Then $EX(r,n,m,k) = EY(r,n,m,k)$ for all $(r,n) \in I$, iff

$$\int_0^1 (F^{-1}(t)-G^{-1}(t))\,(1-t)^{k+(n-r)(m+1)-1}\,g_m^{\,r-1}(t)\,dt$$

$$=\begin{cases} (m+1)^{1-r}\int_0^1 (F^{-1}(t)-G^{-1}(t))\,(1-t)^{k+(n-r)(m+1)-1}\,(1-(1-t)^{m+1})^{r-1}\,dt\,, & m>-1 \\ \int_0^1 (F^{-1}(t)-G^{-1}(t))\,(1-t)^{k-1}\,(\log\frac{1}{1-t})^{r-1}\,dt\,, & m=-1 \end{cases}$$

$$=\begin{cases} (m+1)^{1-r}\int_0^1 (F^{-1}(1-t)-G^{-1}(1-t))\,t^{k-1}\,(t^{m+1})^{n-r}\,(1-t^{m+1})^{r-1}\,dt\,, & m>-1 \\ \int_0^1 (F^{-1}(t)-G^{-1}(t))\,(1-t)^{k-1}\,(\log\frac{1}{1-t})^{r-1}\,dt\,, & m=-1 \end{cases}$$

$= 0$ for all $(r,n) \in I$.

Thus, $$F^{-1}(t) = G^{-1}(t) \quad \text{a.e. on } (0,1),$$

since the corresponding function sequences are complete.

Using Lemma 2.2.1., we have

$$F(x) = G(x) \quad \text{for all } x \in \mathbb{R}.$$

REMARKS 2.2.4.

i) The existence of the moments in 2.2.3. can be ensured globally by results in II.1. If $E|X|^{\alpha} < \infty$ is assumed for a suitable $\alpha > \delta$ (cf. 1.1.3.), $F^{-1}(t)\,(1-t)^{k-1}$ can be integrated.

In the case $k = 1$, only $E|X|$, $E|X|^{\delta} < \infty$ has to be assumed.

ii) In the case $m > -1$, Lemma 2.1.6. and Lemma 2.1.7. contain appropriate sets of indices.

The proof of the theorem indicates that the completeness of a sequence of polynomials $((t^{m+1})^n)_{n \geq r}$ may be used, if $m > -1$ (see 2.1.3. and 2.1.4.).

If $m = -1$, Lemma 2.1.8. may be applied with

$$I = \{ (r,n) \mid \text{for each } r \in \mathbb{N} \text{ choose any } n \geq r \}.$$

iii) If $m = -1$ and $F^{-1}(t)\,(1-t)^{k-1}\,(\log \frac{1}{1-t})^{r_0}$ is $L_{\delta}(0,1)$–integrable for some fixed $r_0 \in \mathbb{N}_0$, the assertion remains valid for

$$I = \{ (r_0 + r,n) \mid \text{for each } r \in \mathbb{N} \text{ choose any } n \geq r_0 + r \}.$$

Using Lemma 2.1.8. the question arises as to whether $\mathbb{N}$ may be replaced by a subset of $\mathbb{N}$. A restriction on a subsequence as appearing in the Müntz–Szász lemma (2.1.2.) is not possible in general; Lin, Huang (1987) present an example.

iv) Applying Corollary 2.1.11., the assertion of the theorem remains valid for $m = -1$ assuming $L_1(0,1)$–integrability with respect to some interval $(0,\delta)$, $0 < \delta < 1$, if $F^{-1}(t)\,(1-t)^{k-2} \in L(0,1)$ is required.

v) Results are also possible in the case $m < -1$. However, it is not permitted to choose a sequence with respect to r or n fixing m and k, because the definition of g OS's requires $k + (n-r)(m+1) \geq 1$.

If the completeness of $(t^{k_j-1})_{j\in\mathbb{N}}$ on $L(0,1)$ is assumed and if $F^{-1}(1-t)$ $\cdot\, (t^{m+1})^{n-r}\,(1-t^{m+1})^{r-1}$ is supposed to be $L(0,1)$–integrable, the sequence of expectations $(\,EX(r,n,m,k_j)\,)_{j\in\mathbb{N}}$ characterizes the distribution function F. Another possibility is to couple some sequences $(k_j)_j$ and $(n_j)_j$.

vi) Similar results may be obtained by considering sequences $(k_j)_j$ or $(m_j)_j$ or combinations of indices with respect to r, n, m or k , if the completeness of the corresponding function sequences is given and if the integrals are well defined.

vii) The theorem may also be formulated in the case of non–identical constants m_i , $i = 1,\dots,n-1$.
Let, e.g., $r' = \max\{r;\, (r,n) \in I\}$; then, assuming $m_1 = \dots = m_{r'-1} = m$, it has to be ensured that the sequence

$$(\,(1-t)^{k+n-r+M_r-1}\, g_m^{r-1}(t)\,)_{(r,n)\in I}$$

is complete on $L(0,1)$.

viii) If F is continuous, generalizations to arbitrary moments $EX^{\nu}(r,n,m,k)$, $\nu \in \mathbb{N}$, follow from 2.2.2.
Assume the existence of moments of order ν, and let the assumptions of Theorem 2.2.3. be valid for $(F^{-1})^{\nu}$ (instead of F^{-1}), then the sequence

$$(\,EX^{\nu}(r,n,m,k)\,)_{(r,n)\in I}$$

characterizes the distribution function F.
In the case of records, this is a generalization of a result of Kirmani, Beg (1984) stated in Lin (1987).

For o OS′s, results of this type are well known and frequently used (see the introduction to the second section). A general version is stated in Galambos (1975).

THEOREM 2.2.5. Let F be a distribution function with an absolutely continuous pseudo–inverse F^{-1} . Let I be a set of pairs $(\nu_1,\nu_2) \in \mathbb{N}^2$ such that the function sequence

$$\begin{cases} \left((t^{m+1})^{\nu_2-\nu_1}(1-t^{m+1})^{\nu_1-1}\right)_{(\nu_1,\nu_2)\in I} & \text{is complete on } L(0,1)\text{ , if } m>-1 \\ \left((\log \frac{1}{1-t})^{\nu_1-1}\right)_{(\nu_1,\nu_2)\in I} & \text{is complete on } L_\delta(0,1)\text{ , } \delta>1\text{, if } m=-1 \end{cases}$$

If for some fixed parameters k and m

$$\begin{cases} (F^{-1})'(t)\,(1-t)^k\,(1-(1-t)^{m+1}) \in L(0,1) & \text{, if } m>-1 \\ (F^{-1})'(t)\,(1-t)^k\,(\log\frac{1}{1-t}) \in L_\delta(0,1) & \text{, if } m=-1 \end{cases}$$

and if the existence of the expectations of Theorem 2.2.3. is ensured for all (r,n) with $(r-1,n-1)\in I$, then the sequence

$$(EX(r,n,m,k)-EX(r-1,n,m,k))_{(r-1,n-1)\in I}$$

characterizes the distribution function F up to a location parameter.

Proof Let $X(r,n,m,k)$ and $Y(r,n,m,k)$ be g OS's based on F and G, respectively. Using Theorem 1.3.1. we have :

$$EX(r,n,m,k)-EX(r-1,n,m,k) = \frac{c_{r-1}}{(r-1)!}\int_0^1 (F^{-1})'(t)\,(1-t)^{k+(n-r)(m+1)}\,g_m^{r-1}(t)\,dt\,.$$

Thus,

$$EX(r,n,m,k)-EX(r-1,n,m,k) = EY(r,n,m,k)-EY(r-1,n,m,k)$$

for all (r,n) satisfying $(r-1,n-1)\in I$ holds, iff

$$\int_0^1 ((F^{-1})'(t)-(G^{-1})'(t))\,(1-t)^{k+(n-r)(m+1)}\,g_m^{r-1}(t)\,dt$$

$$= \begin{cases} (m+1)^{1-r}\int_0^1((F^{-1})'(t)-(G^{-1})'(t))\,(1-t)^{k+(n-r)(m+1)}\,(1-(1-t)^{m+1})^{r-1}\,dt\text{ , } m>-1 \\ \int_0^1((F^{-1})'(t)-(G^{-1})'(t))\,(1-t)^k\,(\log\frac{1}{1-t})^{r-1}\,dt\text{ , } m=-1 \end{cases}$$

$$= \begin{cases} (m+1)^{1-r}\int_0^1((F^{-1})'(1-t)-(G^{-1})'(1-t))\,t^k\,(t^{m+1})^{n-r}\,(1-t^{m+1})^{r-1}\,dt\text{ , } m>-1 \\ \int_0^1((F^{-1})'(t)-(G^{-1})'(t))\,(1-t)^k\,(\log\frac{1}{1-t})^{r-1}\,dt\text{ , } m=-1 \end{cases}$$

$$= \begin{cases} (m+1)^{1-r} \int_0^1 ((F^{-1})'(1-t)-(G^{-1})'(1-t))t^k(1-t^{m+1})(t^{m+1})^{(n-1)-(r-1)}(1-t^{m+1})^{r-2}dt \\ \int_0^1 ((F^{-1})'(t)-(G^{-1})'(t))\,(1-t)^k\,(\log \frac{1}{1-t})\,(\log \frac{1}{1-t})^{r-2}\,dt \end{cases}$$

$= 0$ for all (r,n) with $(r-1,n-1) \in I$.

Since the corresponding function sequences are complete, we have

$$(F^{-1})'(t) = (G^{-1})'(t) \quad \text{a.e. on } (0,1).$$

Thus, there exists a $\mu \in \mathbb{R}$ such that $F^{-1}(t) = G^{-1}(t) + \mu$ a.e. on $(0,1)$.

Using Lemma 2.2.1., the assertion of the theorem follows :

$$F(x+\mu) = G(x) \quad \text{for all } x \in \mathbb{R}.$$

REMARKS 2.2.6.

i) The representation for the difference of expectations used in the proof of 2.2.5. is valid for $r \geq 2$. In some sets of indices (see Lemma 2.1.6.) which guarantee the completeness of $(x^{\nu_1-1}(1-x)^{\nu_2-\nu_1})_{(\nu_1,\nu_2)\in I}$, $\nu_1 = 1$ is needed. That is the reason for the assumption $(r-1,n-1) \in I$.

In particular, this is important in the case $m = -1$ (cf. 2.1.8.), since the set of exponents of $\log \frac{1}{1-t}$ equals $\mathbb{N}_0$.

ii) In the theorem, $(F^{-1})'(t)\,(1-t)^k\,(1-(1-t)^{m+1})$ is assumed to be $L(0,1)$–integrable and $(F^{-1})'(t)\,(1-t)^k\,(\log \frac{1}{1-t})$ to be $L_\delta(0,1)$–integrable, respectively. These conditions are not really strong; e.g., assuming the existence of $EX(2,2,m,k+1)$ and $EX^2(3,n,-1,2k+1)$ $(\delta < 2)$ respectively, is sufficient.

iii) In the theorem, only the existence of those expectations is required which enable us to apply Theorem 2.2.3. However, the existence can be ensured globally using results stated in the first section.

iv) Cf. Remark 2.2.4.ii) .

v) If $m = -1$ and $F^{-1}(t)\,(1-t)^k\,(\log \frac{1}{1-t})^{r_0+1}$ is assumed to be $L_\delta(0,1)$–integrable ($\delta > 1$) for some fixed $r_0 \in \mathbb{N}_0$, then the assertion of the theorem remains valid for

$I = \{\, (r_0 + r, n) \mid$ for each $r \in \mathbb{N}$ choose some $n \geq r_0 + r \,\}$ (cf. Lemma 2.1.8.).

vi) Cf. Remark 2.2.4.iv) .

vii) For $m < -1$ we refer to Remark 2.2.4.v).

viii) Remark 2.2.4.vi) with respect to Theorem 2.2.3. may also lead to some extensions here.

ix) As in Theorem 2.2.3., generalizations to higher moments are possible using Theorem 1.3.1.

Then, we obtain

$$(F^{-1}(t))^{\alpha-1}\,(F^{-1})'(t) = (G^{-1}(t))^{\alpha-1}\,(G^{-1})'(t) \quad \text{a.e. on } (0,1)$$

in the proof of the theorem.

Hence, for $\alpha \neq 0$ we have

$$((F^{-1}(t))^{\alpha} + c_1)' = ((G^{-1}(t))^{\alpha} + c_2)' \quad \text{a.e. on } (0,1);$$

i.e., there exists $\mu \in \mathbb{R}$ with $(F^{-1}(t))^{\alpha} = (G^{-1}(t))^{\alpha} + \mu$ a.e. on $(0,1)$.

x) In Theorem 2.2.5., the assumption of equal constants $m_1,\ldots,m_{n-1}$ may be dropped under suitable conditions (cf. Theorem 2.2.3. and 2.2.4.vii)).
The assertion of 2.2.5. is stated by Saleh (1976) and Lin (1988b) in the case of o OS′s and by Gupta (1984) for record values (using higher differences).

Characterizations of probability distributions by sequences of moments, results on the completeness of certain function sequences and many other facts and investigations may be found in Arnold, Meeden (1975) and in a series of papers by Huang, Hwang and Lin (Huang 1975, 1989, Hwang 1978, 1983, Lin 1984, 1986, 1987, 1988b, 1989a, Huang, Hwang 1975, Hwang, Lin 1984a,b, 1987, Lin, Huang 1987). The articles by Hwang, Lin (1984a), Huang (1989) and Lin (1989a) are reviews on this topic.

Other applications of results on completeness and characterizations of probability distributions by moments of o OS's and record values can be found, e.g., in a paper of Huang (1974b) in connection with Ahsanullah, Rahman (1972) and Huang (1974a) (cf. Kamps 1990b, 1992b); moreover, we refer to Chapter III and the cited literature therein and to Galambos, Kotz (1978), Azlarov, Volodin (1986).

Chapter III

Recurrence Relations for Moments of Generalized Order Statistics and Characterizations of Distributions

Recurrence relations and identities for moments of o OS's have been extensively investigated in the literature. Examples of explicit expressions for moments are given in Chapter II and, obviously, they may lead to a high numerical effort and to problems concerning the accuracy of computational results. Therefore, recurrence relations are a very useful tool. Here, relations between moments are termed 'recurrence relations' throughout; but we do not use this term in the strict sense. I.e., it is not necessarily possible to reconstruct a whole system of moments via some given set of moments which may be suggested by the parametrization of a certain identity. In contrast to o OS's, recurrence relations for record values have been considered in the literature only marginally. Some relations are obtained in connection with characterizations of distributions by means of inequalities (see Chapter IV, Lin 1988a, Gajek, Gather 1991, Kamps 1991a). Another example is shown in Lin (1988b). Thus, we give some examples in the special case $m = -1$. In principle, the identities can be classified with respect to their validity for arbitrary or special distributions.

For the sake of a simplified presentation, we consider g OS's $X(r,n,m,k)$ only; i.e., we assume $m_1 = ... = m_{n-1} = m$ throughout in this chapter.

1. Recurrence Relations for Arbitrary Distributions

The most important recurrence relation for moments of o OS's from arbitrary distributions is given by Cole (1951) in the continuous case and it is frequently used:

$$(n-r)\, EX^{\alpha}_{r,n} + r\, EX^{\alpha}_{r+1,n} = n\, EX^{\alpha}_{r,n-1}, \quad 1 \leq r \leq n-1 .$$

Melnick (1964) proves it in the discrete case. Using the integral representation of a moment via the pseudo–inverse of some underlying distribution function F (cf. II.1.1.1.), the assertion is seen to hold true for arbitrary distributions, obviously (see I.3.2.2.). Moreover, the assumption of independence of the underlying identically distributed random variables $X_1,\dots,X_n$ can be weakened; assuming exchangeable random variables turns out to be sufficient (David, Joshi 1968).

This identity has the following interpretation :

If all moments of o OS's of order α are known in a sample of size $n-1$ and if, moreover, $EX^{\alpha}_{i,n}$ is known for any i, $1 \le i \le n$, then all moments of order α in a sample of size n can be computed.

In connection with the well–known result of Hoeffding (1953), this relation is mentioned in Section II.2. on the completeness of function sequences and characterizations of distributions by moments.
Identities for arbitrary distributions are shown in David (1981), Arnold, Balakrishnan (1989), Balakrishnan, Cohen (1991), Arnold, Balakrishnan, Nagaraja (1992) and in the detailed review by Malik, Balakrishnan, Ahmed (1988).
Moreover, we point out the important results on recurrence relations and identities for the moments and distribution functions of o OS's from dependent r.v.'s which can be found, e.g., in Young (1967), David, Joshi (1968), Balakrishnan (1987), Sathe, Dixit (1990), Balasubramanian, Bapat (1991), Balakrishnan, Bendre, Malik (1992), Balasubramanian, Balakrishnan (1993) and David (1993a).

The above identity can be extended to g OS's and is stated in the following corollary. In the special case of record values ($m = -1$) the identities become trivial, however.

Corollary 1.1. For moments of g OS's with $m_1 = \dots = m_{n-1} = m$ and $1 \le r \le n-1$ the following identities are valid (subject to the existence of the expectations) :

i)
$$(k+(n-r-1)(m+1))\, EX^{\alpha}(r,n,m,k) + r(m+1)\, EX^{\alpha}(r+1,n,m,k)$$
$$= (k+(n-1)(m+1))\, EX^{\alpha}(r,n-1,m,k)\,,$$

ii)
$$(k+(n-r-1)(m+1)) \left(EX^{\alpha}(r+1,n,m,k) - EX^{\alpha}(r,n,m,k)\right)$$
$$= (k+(n-1)(m+1)) \left(EX^{\alpha}(r+1,n,m,k) - EX^{\alpha}(r,n-1,m,k)\right),$$

iii) $$(k+(n-1)(m+1))\left(EX^{\alpha}(r,n,m,k) - EX^{\alpha}(r,n-1,m,k)\right)$$
$$= r(m+1)\left(EX^{\alpha}(r,n,m,k) - EX^{\alpha}(r+1,n,m,k)\right).$$

Proof The assertions immediately follow from Lemma I.3.2.2.

Using the identities of this corollary, the relations shown in this chapter can be modified. For o OS's this is pointed out in Kamps (1991b).

2. Recurrence Relations for Specific Distributions

A variety of results on recurrence relations for moments of o OS's from specific distributions is found in the literature. The articles of Malik (1966, 1967) und Joshi (1977) have previously been cited and we refer to Barnett (1966) and Khan, Khan (1983, 1987), e.g. Usually, explicit expressions for moments are the starting point to derive identities by means of structural properties. In the case of the logistic distribution, however, this way to proceed turns out to be difficult. Gupta, Shah (1965) are concerned with moments of o OS's from the logistic distribution and give a representation via double sums using products of binomial coefficients, Bernoulli numbers and Stirling numbers of the first kind; alternatively, digamma und polygamma functions have to be evaluated. Shah (1970) points out the validity of the relation $F(x)\ (1-F(x)) = f(x)$ for logistic distributions which immediately leads to an important identity. In works of Khan, Yaqub, Parvez (1983) and Lin (1988b) we find a more systematic treatment. Once having derived a certain representation for the difference of moments of successive o OS's, they put in different distribution functions to obtain similar recurrence relations.
Identities in the case of special distributions are shown in Govindarajulu (1963), David (1981), Arnold, Balakrishnan (1989), Balakrishnan, Cohen (1991) and Arnold, Balakrishnan, Nagaraja (1992). In particular, we refer to the very detailed survey article of Balakrishnan, Malik, Ahmed (1988) and to Khan, Yaqub, Parvez (1983) where truncated distributions are considered. There are also many results concerning product moments. We do not give related results here, but refer to the above cited literature and, e.g., to Joshi, Balakrishnan (1982) and Lin (1989b).

Summarizing, the following may be noted in connection with the use and application of recurrence relations and identities for moments of o OS′s :
In comparison with direct computations via explicit expressions, we have a reduced numerical effort and a raised accuracy, in general. Higher moments or moments with respect to a large sample may be derived in a simple manner. Moreover, the identities can be used to check numerical results and, as a by–product, we obtain combinatorial identities. Some more aspects are important and are discussed in the sequel. Identities for moments of OS′s may provide an insight into the structure of certain classes of distributions and new characterizations of distributions arise. By this, we are able to avoid separate investigations when defining appropriate families of distributions. Proceeding from explicit expressions, the known results often are not as general as possible with respect to their parametrization. Moreover, similarly structured identities, and by this relationships to other distributions, remain hidden.

There are a few characterization results for distributions by recurrence relations (cf. Azlarov, Volodin 1986, Lin 1988b). Lin (1988b) considers similarly structured characterizing recurrence relations for uniform, Pareto, exponential and logistic distributions :

$$EX_{r,n}^{\alpha} - EX_{r-1,n}^{\alpha} = \alpha\, c(r,n,p,q)\, EX_{r+p,n+q}^{\alpha-1}$$

with integers p,q and certain constants $c(r,n,p,q)$.

In this context, a characterization set–up can become a method. Choosing a parametrized recurrence relation as the starting point, a characterization set–up may lead to a corresponding parametrized family of distributions (see Theorem 2.1.5.). In a second step, the validity of the identity for the elements of this class may be shown easily and under mild conditions. Hence, relationships between distributions become obvious, isolated results known from the literature can be subsumed and integrated within a general framework, well known identities can be generalized with respect to the parametrization of the underlying distributions and to moments of non–integral orders and new relations are found.

This method is demonstrated for three classes of distributions. First, we establish some parametrized recurrence relation together with the corresponding characterization result in the case of o OS′s for a class of distributions including, e.g., exponential, power function, Pareto and logistic distributions (cf. Kamps 1991b). The results are then transferred to g OS′s. In this identity moments of orders α and $\alpha - 1$ are involved. We then state a more general approach leading to recursions for distributions with an additional parameter

(e.g. Burr XII distributions, see Kamps 1992a). This new class of distributions includes the former one. We give the corresponding result for g OS's and show some examples in the special cases of o OS's and record values. The definition of the third class of distributions is motivated by a recurrence relation of Barnett (1966) for the moments of OS's from a Cauchy distribution which is generalized by Khan, Yaqub, Parvez (1983) to doubly truncated Cauchy distributions. Again, the consideration in the case of o OS's (see Kamps 1990a) is followed by its generalization to g OS's. At the end of the second section we gather together some other identities for specific distributions.
In order to characterize distributions by means of recurrence relations, a strong assumption is made: the validity of some identity is required with respect to a sequence of indices. Too, Lin (1989) show identities for moments of o OS's and record values of first and second orders which turn out to determine certain power function and Weibull distributions without any further assumptions. A generalization of this result to g OS's is stated in the third section.

2.1. The Class $\mathcal{F}$ of Distributions

Proceeding from the parametrized recurrence relation for moments of o OS's in Theorem 2.1.4., the characterization set-up of Theorem 2.1.5. leads to the class $\mathcal{F}$ of distribution functions defined by

2.1.1. $$(F^{-1})'(t) = \frac{1}{d}\, t^{p}\, (1-t)^{q-p-1} \quad \text{a.e. on } (0\,,\,1)$$

with constants $d > 0$ and $p,q \in \mathbb{Z}$ (see Kamps 1991b).

Examples 2.1.2.

To determine F and F^{-1} from 2.1.1. respectively, we distinguish nine cases with respect to p and q (see e.g. Gröbner, Hofreiter 1, 1961), $c \in \mathbb{R}$:

i) $p = 0\,,\, q = 0$:

$F^{-1}(t) = \frac{1}{d} \log \frac{1}{1-t} + c$; hence

$F(x) = 1 - \exp\{-d(x-c)\}\,,\; x \in (c\,,\,\infty)$

(exponential distributions).

ii) $p = 0, q \neq 0$:

$$F^{-1}(t) = -\frac{1}{dq}(1-t)^q + c; \text{ hence}$$

$$F(x) = 1 - (dq(c-x))^{1/q}, \quad \begin{cases} x \in (c - \frac{1}{dq}, c), q > 0 \\ x \in (c - \frac{1}{dq}, \infty), q < 0 \end{cases}$$

($q < 0$: Pareto, special Burr XII– (i.e. Lomax–) distributions (e.g. Tadikamalla 1980); $q > 0$, $d = \frac{1}{qc}$: special Pearson–type 1 distributions, special case of a generalized J shaped beta distribution (cf. Houchens 1984, p 70)).

iii) $p \neq -1, q = p+1$:

$$F^{-1}(t) = \frac{1}{dq} t^q + c; \text{ hence}$$

$$F(x) = (dq(x-c))^{1/q}, \quad \begin{cases} x \in (c, c + \frac{1}{dq}), q > 0 \\ x \in (-\infty, c + \frac{1}{dq}), q < 0 \end{cases}$$

($q > 0$: power function distributions).

iv) $p = -1, q = 0$:

$$F^{-1}(t) = \frac{1}{d} \log t + c; \text{ hence } F(x) = \exp\{d(x-c)\}, \ x \in (-\infty, c).$$

v) $p \in \mathbb{N}, q \notin \{0,1,\dots,p\}$:

$$F^{-1}(t) = -\frac{1}{d} \sum_{j=0}^{p} (-1)^j \binom{p}{j} \frac{1}{q-p+j} (1-t)^{q-p+j} + c.$$

vi) $p \notin \{-1,\dots,-(q-p)\}, q-p \geq 2$:

$$F^{-1}(t) = \frac{1}{d} \sum_{j=0}^{q-p-1} (-1)^j \binom{q-p-1}{j} \frac{1}{p+j+1} t^{p+j+1} + c.$$

vii) $p \in \mathbb{N}, q \in \{0,1,\dots,p\}$:

$$F^{-1}(t) = -\frac{1}{d} \sum_{j=0,\ j \neq p-q}^{p} (-1)^j \binom{p}{j} \frac{1}{q-p+j} (1-t)^{q-p+j} + \frac{1}{d}(-1)^q \binom{p}{p-q} \log(1-t) + c.$$

viii) $p \in \{-1,\dots,-(q-p)\}, q-p \geq 2$:

$$F^{-1}(t) = \frac{1}{d} \sum_{j=0,\ j \neq -(p+1)}^{q-p-1} (-1)^j \binom{q-p-1}{j} \frac{1}{p+j+1} t^{p+j+1} + \frac{1}{d}(-1)^{p+1} \binom{q-p-1}{-p-1} \log t + c.$$

ix) $p < 0\,,\ q < p+1$:

$$F^{-1}(t) = \frac{1}{d}\binom{-q-1}{p-q}\log\frac{t}{1-t} + \frac{1}{d}\sum_{j=0,\ j\neq p-q}^{-q-1}\binom{-q-1}{j}\frac{1}{p-q-j}\left(\frac{t}{1-t}\right)^{p-q-j} + c$$

($p = -1\,,\ q = -1$: logistic distributions).

EXAMPLES : ad vii) $p = 1\,,\ q = 1:\ F^{-1}(t) = \frac{1}{d}(1-t-\log(1-t)) + c\,;$

ad viii) $p = -1\,,\ q = 1:\ F^{-1}(t) = -\frac{1}{d}(t-\log t) + c\,.$

REMARK 2.1.3. A subclass of the presented family may be described in this way : Consider those parameter combinations with respect to p and q where, for some r, n with $1 \le r \le n$, it is permitted to put $p = r-1$ and $q = n$. Then the distribution function F is determined such that the distribution function of the r–th o OS based on F is some uniform distribution; to be more precise :

$$F^{X_{r,n}}(x) = d\, r \binom{n}{r} (x-c)$$

with constants c and $d > 0$.

PROOF Since $(F^{-1})'(t) = \frac{1}{d} t^p (1-t)^{q-p-1}$ and $F^{-1}(x) = \frac{1}{d}\int_0^x t^p (1-t)^{q-p-1}\,dt + c$, we observe $x = \frac{1}{d}\left(r\binom{n}{r}\right)^{-1} F^{X_{r,n}}(x) + c$ with $p = r-1\,,\ q = n\,.$

For any distribution function belonging to $\mathcal{F}$, we state a corresponding parametrized recurrence relation for moments of o OS's containing the cited results of Lin (1988b). The constant c(r,n,p,q) appearing in the equations turns out to be the expectation of a certain spacing.

Here, we restrict ourselves to positive moments. In the case of negative moments the results remain valid, if the support of F is contained in the positive real line. If negative integer moments are considered, the support of F may alternatively be contained in the positive or in the negative real line.

THEOREM 2.1.4. Let the appearing OS's be based on some distribution function $F \in \mathcal{F}$, let $\alpha \geq 1$ be a constant and $F^{-1}(0) \geq 0$, if $\alpha \notin \mathbb{N}$.

Then for all $r, n \in \mathbb{N}$, $2 \leq r \leq n$, satisfying

$$1 \leq r+p \leq n+q$$

and
$$-\infty < EX^{\alpha}_{r,n}, EX^{\alpha}_{r-1,n}, EX^{\alpha-1}_{r+p,n+q} < \infty,$$

the identity

$$EX^{\alpha}_{r,n} - EX^{\alpha}_{r-1,n} = \alpha\, c(r,n,p,q)\, EX^{\alpha-1}_{r+p,n+q}$$

is valid where the constant $c(r,n,p,q)$ is given by

$$c(r,n,p,q) = EX_{r,n} - EX_{r-1,n} = \frac{1}{d} \frac{\binom{n}{r-1}}{(r+p)\binom{n+q}{r+p}}.$$

PROOF Since

$$EX^{\alpha}_{r,n} = r\binom{n}{r} \int_0^1 (F^{-1}(t))^{\alpha}\, t^{r-1} (1-t)^{n-r}\, dt,$$

we obtain

$$EX^{\alpha}_{r,n} - EX^{\alpha}_{r-1,n} = \alpha \binom{n}{r-1} \int_0^1 (F^{-1}(t))^{\alpha-1} (F^{-1})'(t)\, t^{r-1} (1-t)^{n-r+1}\, dt$$

via integration by parts (see Khan, Yaqub, Parvez 1983, Lin 1988b and Theorem II.1.3.1. for g OS's).

Putting in the representations of $(F^{-1})'(t)$ and $c(r,n,p,q)$, the assertion follows.

In particular, we have

$$EX_{r,n} - EX_{r-1,n} = c(r,n,p,q),$$

since (cf. David, Groeneveld 1982)

$$\begin{aligned} EX_{r,n} - EX_{r-1,n} &= \binom{n}{r-1} \int_0^1 (F^{-1})'(t)\, t^{r-1} (1-t)^{n-r+1}\, dt \\ &= \frac{1}{d} \binom{n}{r-1} \int_0^1 t^{r+p-1} (1-t)^{n-r+q-p}\, dt \\ &= \frac{1}{d} \frac{\binom{n}{r-1}}{(r+p)\binom{n+q}{r+p}} = c(r,n,p,q). \end{aligned}$$

The following table shows the distribution functions and the constants $c(r,n,p,q)$ with respect to some parameter combinations of p and q.

p	q	F(x)	x ∈	Distribution	d·c(r,n,p,q)
0	0	$1-\exp\{-d(x-c)\}$	$(c\,,\,\infty)$	exponential	$\frac{1}{n-r+1}$
0	> 0	$1-(dq(c-x))^{1/q}$	$(c-\frac{1}{dq},\, c)$	Pearson I	$\frac{n!\,(n-r+q)!}{(n+q)!\,(n-r+1)!}$
0	< 0	$1-(dq(c-x))^{1/q}$	$(c-\frac{1}{dq},\, \infty)$	Pareto, Lomax	$\frac{n!\,(n-r+q)!}{(n+q)!\,(n-r+1)!}$
−1	0	$\exp\{d(x-c)\}$	$(-\infty\,,\, c)$		$\frac{1}{r-1}$
>−1	p+1	$(dq(x-c))^{1/q}$	$(c\,,\, c+\frac{1}{dq})$	power function	$\frac{n!\,(r+p-1)!}{(n+p+1)!\,(r-1)!}$
<−1	p+1	$(dq(x-c))^{1/q}$	$(-\infty,\, c+\frac{1}{dq})$		$\frac{n!\,(r+p-1)!}{(n+p+1)!\,(r-1)!}$
−1	−1	$(1+\exp\{-d(x-c)\})^{-1}$	$(-\infty\,,\,\infty)$	logistic	$\frac{n}{(r-1)(n-r+1)}$

A modification of this recurrence relation, and by this a modification of the following characterization theorem too, is obtained observing that the equation

$$EX^{\alpha}_{r,n} - EX^{\alpha}_{r-1,n} = \frac{n}{n-r+1}(EX^{\alpha}_{r,n} - EX^{\alpha}_{r-1,n-1})\,,\; 2 \le r \le n\,,$$

is valid for an arbitrary distribution (see Corollary 1.1.).

Under the additional assumption $F^{-1}(0) = 0$, the representation of $EX^{\alpha}_{r,n} - EX^{\alpha}_{r-1,n}$ remains valid in the case $r = 1$ with $EX^{\alpha}_{0,n} = 0$.

Assuming the validity of the recurrence relation in Theorem 2.1.4. for an appropriate sequence of pairs of indices $((r_j,n_j))_{j\in\mathbb{N}}$, the correspondingly parametrized distribution function in the family $\mathcal{F}$ can be characterized. The sequence has to be chosen such that the sequence of polynomials

$$\{t^{r_j-1}(1-t)^{n_j-r_j+1}\}_j \quad \left[\text{or } \{(1-t)^{n_j}\}_j\right]$$

is complete on the space L(0 , 1). Appropriate sequences may be found in Chapter II.2. Such a characterization of the exponential distribution is given by Azlarov, Volodin (1986) in the case $\alpha = 2$. Lin (1988b) obtains characterizations of exponential, uniform, Pareto and logistic distributions applying the Müntz–Szász Lemma (see II.2.1.2.).
These results are contained in

THEOREM 2.1.5. Let the OS's be based on the distribution function F, let F^{-1} be absolutely continuous on the interval (0 , 1) and let $\alpha \geq 1$ be a constant with

$$\begin{cases} |\{t \in (0,1);\, F^{-1}(t) = 0\}| \in \{0,1\}, & \alpha \in \mathbb{N}^{\geq 2} \\ F^{-1}(t) > 0 \text{ for all } t \in (0,1), & \alpha \notin \mathbb{N} \end{cases}$$

Moreover, let integers $p, q \in \mathbb{Z}$ and a sequence $((r_j,n_j))_{j\in\mathbb{N}}$ according to the above remark are given satisfying the conditions

$$2 \leq r_j \leq n_j, \quad 1 \leq r_j + p \leq n_j + q$$

and

$$-\infty < EX^{\alpha}_{r_j,n_j},\ EX^{\alpha}_{r_j-1,n_j},\ EX^{\alpha-1}_{r_j+p,n_j+q} < \infty$$

for all $j \in \mathbb{N}$.
Then for some constant $d > 0$ we have

$$(F^{-1})'(t) = \tfrac{1}{d}\, t^{p}(1-t)^{q-p-1} \quad \text{a.e. on } (0\,,1)\,,$$

iff

$$EX^{\alpha}_{r_j,n_j} - EX^{\alpha}_{r_j-1,n_j} = \alpha\, c(r_j,n_j,p,q)\, EX^{\alpha-1}_{r_j+p,n_j+q}$$

with

$$c(r_j,n_j,p,q) = \frac{1}{d}\, \frac{\binom{n_j}{r_j-1}}{(r_j+p)\binom{n_j+q}{r_j+p}}$$

for all elements of the sequence $((r_j,n_j))_{j\in\mathbb{N}}$ and assuming

$$(F^{-1}(t))^{\alpha-1}\left[d\,(F^{-1})'(t) - t^{p}(1-t)^{q-p-1}\right] \in L(0\,,1)\,.$$

PROOF Since $\quad EX^{\alpha}_{r_j,n_j} - EX^{\alpha}_{r_j-1,n_j} = \alpha\, c(r_j,n_j,p,q)\, EX^{\alpha-1}_{r_j+p,n_j+q}$

and $\quad \int_0^1 (F^{-1}(t))^{\alpha-1}\, t^{r_j-1}\, (1-t)^{n_j-r_j+1} \left[d\,(F^{-1})'(t) - t^p\,(1-t)^{q-p-1}\right] dt = 0$

are equivalent, we conclude

$$d\,(F^{-1})'(t) - t^p\,(1-t)^{q-p-1} = 0 \quad \text{a.e. on } (0\,,1)\,.$$

REMARKS

i) If $r \in \mathbb{N}$ is fixed (e.g. using sequences according to the lemma of Müntz, Szász, cf. II.2.1.2.) the constant d may depend on r.

ii) Applying another complete function sequence (see II.2.1.), an appropriately modified integrability condition is required.

iii) The characterization result itself is less important in view of the strong assumptions. However, this approach leads to a class of distributions the elements of which are related by a similar recurrence relation for moments of o OS's. In this regard, the characterization set-up works as a method for a systematic treatment of recurrence relations.

Now, we transfer the results for o OS's to the case of g OS's. For that purpose, we replace powers of t by powers of the function g_m and we replace $q-p-1$ by $(q-p)(m+1)+s-1$ with some $s \in \mathbb{R}$ in the definition of the family of distributions.

THEOREM 2.1.6. Let $X(r,n,m,k)$, $X(r-1,n,m,k)$ and $X(r+p,n+q,m,k+s)$, $r \geq 2$, be g OS's based on the distribution function F given by

$$(F^{-1})'(t) = \tfrac{1}{d}\,(1-t)^{(q-p)(m+1)+s-1}\, g_m^p(t) \quad \text{a.e. on } (0,1)$$

with $d > 0$, $p,q \in \mathbb{Z}$, $m \in \mathbb{R}$, $s \geq 1-k$ and let $\alpha \neq 0$ be a constant such that

$$-\infty < EX^{\alpha}(r,n,m,k)\,,\ EX^{\alpha}(r-1,n,m,k)\,,\ EX^{\alpha-1}(r+p,n+q,m,k+s) < \infty\,.$$

Then we obtain the recurrence relation

$$EX^{\alpha}(r,n,m,k) - EX^{\alpha}(r-1,n,m,k) = \alpha\, C\, EX^{\alpha-1}(r+p,n+q,m,k+s)$$

with $$C = \frac{1}{d}\,\frac{c_{r-2}(n\,,k)}{c_{r+p-1}(n+q,k+s)}\,\frac{(r+p-1)!}{(r-1)!} = EX(r,n,m,k) - EX(r-1,n,m,k)\,,$$

if the expectations exist.

PROOF Using Theorem II.1.3.1. and the representation of $(F^{-1})'(t)$ we find :

$EX^{\alpha}(r,n,m,k) - EX^{\alpha}(r-1,n,m,k)$

$$= \frac{\alpha}{d}\,\frac{c_{r-2}(n,k)}{(r-1)!}\int_0^1 (F^{-1}(t))^{\alpha-1}\,(1-t)^{k+s+(n+q-(r+p))(m+1)-1}\,g_m^{\,r+p-1}(t)\,dt$$

$$= \frac{\alpha}{d}\,\frac{c_{r-2}(n\,,k)}{c_{r+p-1}(n+q,k+s)}\,\frac{(r+p-1)!}{(r-1)!}\,EX^{\alpha-1}(r+p,n+q,m,k+s)\,.$$

Putting $\alpha = 1$, we obtain the representation of $EX(r,n,m,k) - EX(r-1,n,m,k)$ and the assertion follows.

EXAMPLES 2.1.7. In the case $m \neq -1$, we obtain representations of the pseudo–inverse F^{-1} as certain sums, again.
The special case $m = -1$ leads to recurrence relations for, e.g., Weibull distributions.
If $m = -1$, the constant C simplifies to

$$C = \frac{1}{d}\,\frac{k^{r-1}}{(k+s)^{r+p}}\,\frac{(r+p-1)!}{(r-1)!}\,.$$

Putting $p = s = 0$, we find a recursion for exponential distributions with

$$F(x) = 1 - \exp\{-d(x-c)\}\;,\; x \in (c,\infty),$$

(cf. Lemma 2.4.3.iv)). Choosing $p = 0$ and $s < 0$ leads to formulae for Pareto and Lomax distributions (cf. 2.1.2.ii)). Putting $p = 0$ and $s > 0$, we find identities for distribution functions

$$F(x) = 1 - (ds(c-x))^{1/s}\;,\; x \in (c-\tfrac{1}{ds}\,,\, c)\,.$$

In general, $$(F^{-1})'(t) = \tfrac{1}{d}\,(1-t)^{s-1}\,(\log\tfrac{1}{1-t})^p \quad \text{a.e. on } (0,1)$$

with $p \in \mathbb{Z}$, $s \geq 1-k$, is the representation of the derivative of the pseudo–inverse in the case $m = -1$.

Applying certain integration formulae (see Gröbner, Hofreiter 1, 1961, p 111), $F^{-1}(1-t)$ can be explicitly expressed for special choices of p and s .

For $s = 0$, $p \neq -1$ we find distributions with

$$F(x) = 1 - \exp\{-(d(p+1)(x-c))^{1/(p+1)}\}, \quad \begin{cases} x \in (c,\infty), & p \in \mathbb{N}_0 \\ x \in (-\infty, c), & p < -1 \end{cases};$$

i.e. Weibull distributions, if $p \in \mathbb{N}_0$.

By analogy with the remarks on o OS's, Corollary 1.1. can be used to modify the assertion of the theorem.

In the case of o OS's it is shown in Theorem 2.1.5. that, fixing some parameters, the recurrence relation is characteristic to a corresponding distribution. Applying Theorem II.1.3.1. and results in Section II.2.1., analogous characterization results may also be obtained for g OS's.

2.2. A Class of Transformed Distributions belonging to $\mathcal{F}$

In the previous section we considered recurrence relations for moments of o OS's of the following type : $\quad EX^{\alpha}_{r,n} - EX^{\alpha}_{r-1,n} = \alpha\, c(r,n,p,q)\, EX^{\alpha-1}_{r+p,n+q}$;

i.e., the order of the moment on the r.h.s. is reduced by 1 . This structure is now generalized to moments of order $\alpha + \beta$ on the l.h.s. and a moment of order α on the r.h.s. of the identity, choosing $\alpha \in \mathbb{R}$, $\beta \geq 0$ and $\alpha + \beta \neq 0$. Recurrence relations of this type are shown in Khan, Yaqub, Parvez (1983) for Weibull distributions and in Khan, Khan (1987) for Burr XII distributions. Theorem 2.1.6. is contained in the following result putting $\beta = 1$. Moreover, an identity for expectations of general functions of o OS's is considered and discussed in Kamps, Mattner (1993). Pareto and power function distributions are examples in the case $\beta = 1$ (see 2.1.2. and Lin 1988b) and we find other recurrence relations for moments of o OS's if $\beta \neq 1$. In particular, the case $\beta = 0$ is included yielding a relation between three (or two, respectively) moments of the same order. Putting $p = q = 0$ or $p = -1$, $q = 0$, we obtain identities involving two moments which are valid in the case of certain Pareto and power function distributions and which are stated in Malik (1966, 1967).

We derive a version for g OS's and we give examples in the case of o OS's ($m = 0$) and records ($m = -1$). These special cases are considered in Kamps (1992a); moments of records from Pareto, power function, Burr XII and Weibull distributions obey similar recurrence relations. Again, characterization results may be derived under suitable conditions; we refer to the remarks in Section 2.1.

THEOREM 2.2.1. Let $\alpha \in \mathbb{R}$, $\beta \geq 0$ with $\alpha + \beta \neq 0$, let $X(r,n,m,k)$, $X(r-1,n,m,k)$ and $X(r+p,n+q,m,k+s)$ be g OS's with $2 \leq r \leq n$, $p,q \in \mathbb{Z}$, $s \in \mathbb{R}$, $1 \leq r+p \leq n+q$, $k+s \geq 1$, and let h be a function on (0,1) given by

$$\frac{d}{dt}\, h(t) = \frac{1}{d}\,(1-t)^{(q-p)(m+1)+s-1}\, g_m^p(t) \quad \text{a.e. on } (0,1),$$

$d > 0$, such that the quantity $(\beta\, h(t))^{1/\beta}$ is well defined, if $\beta > 0$.

If the underlying distribution function of the g OS's is expressible as

$$F^{-1}(t) = \begin{cases} \exp\{h(t)\} & , \beta = 0 \\ (\beta\, h(t))^{1/\beta} & , \beta > 0 \end{cases}$$

with $\quad -\infty < EX^{\alpha+\beta}(r,n,m,k)\,,\ EX^{\alpha+\beta}(r-1,n,m,k)\,,\ EX^{\alpha}(r+p,n+q,m,k+s) < \infty$,

then we have the following recurrence relation :

$$EX^{\alpha+\beta}(r,n,m,k) - EX^{\alpha+\beta}(r-1,n,m,k) = (\alpha+\beta)\, C\, EX^{\alpha}(r+p,n+q,m,k+s)$$

with
$$C = \frac{1}{d}\,\frac{c_{r-2}(n,k)}{c_{r+p-1}(n+q,k+s)}\,\frac{(r+p-1)!}{(r-1)!}\,.$$

PROOF If $\beta = 0$, we have $(F^{-1})'(t) = \exp\{h(t)\}\, h'(t) = F^{-1}(t)\, h'(t)$ and if $\beta > 0$, we find $(F^{-1}(t))^{\beta-1}\,(F^{-1})'(t) = h'(t)$.

Thus, using Theorem II.1.3.1. and the representation of $h'(t)$, the assertion follows by analogy with Theorem 2.1.4.

REMARK The constant C is equal to the expectation of some difference of successive g OS's (subject to existence); to be more precise :

If the expected values $EY(r,n,m,k)$ and $EY(r-1,n,m,k)$ of g OS's with an underlying distribution function G exist where $G^{-1}(t) = h(t)$, $t \in (0,1)$ (with h as in Theorem 2.2.1.), we observe (cf. Theorem 2.1.4.) : $C = EY(r,n,m,k) - EY(r-1,n,m,k)$.

EXAMPLES 2.2.2. in the case of o OS's ($m = 0$, $k = 1$ and $s = 0$):

i) $p = 0 , q = 0$:

$$h(t) = \frac{1}{d} \log \frac{1}{1-t} + c ,$$

$\beta = 0 , c \in \mathbb{R}$:

$$F(x) = 1 - e^{cd} x^{-d} , \quad x \in (e^c,\infty) ;$$

i.e. Pareto distributions (Malik 1966) ;

$\beta = 1 \vee (\beta > 0 \wedge c \geq 0)$:

$$F(x) = 1 - \exp\{-d(\tfrac{1}{\beta} x^{\beta} - c)\} , \quad x \in ((\beta c)^{1/\beta},\infty) ;$$

i.e. Weibull distributions (Khan, Yaqub, Parvez 1983), in particular, Rayleigh distributions (Balakrishnan, Malik, Ahmed 1988) and exponential distributions ($\beta = 1$) (Joshi 1977, Khan, Yaqub, Parvez 1983, Azlarov, Volodin 1986, Balakrishnan, Malik, Ahmed 1988, Lin 1988b);

ii) $p = 0 , q \neq 0$:

$$h(t) = -\frac{1}{dq} (1-t)^q + c ,$$

$\beta = 0 , c \in \mathbb{R}$:

$$F(x) = 1 - (dq(c - \log x))^{1/q} , \quad \begin{cases} x \in (e^{c-1/(dq)},e^c) , & q > 0 \\ x \in (e^{c-1/(dq)},\infty) , & q < 0 \end{cases} ;$$

i.e. reflected log power function distributions ($q > 0$), log Pareto distributions ($q < 0$);

$\beta = 1 \vee (\beta > 0 \wedge c \geq \frac{1}{dq})$:

$$F(x) = 1 - (dq(c - \tfrac{1}{\beta} x^{\beta}))^{1/q} , \quad \begin{cases} x \in ((\beta(c-\frac{1}{dq}))^{1/\beta},(\beta c)^{1/\beta}), & q > 0 \\ x \in ((\beta(c-\frac{1}{dq}))^{1/\beta},\infty) & , q < 0 \end{cases} ;$$

i.e. special Pearson–type–I distributions ($\beta = 1$), power function distributions ($\beta > 0$, $q = 1$, $c = 1/d$) ($\beta = 1$ in Lin 1988b); $q < 0$: Burr XII distributions (Khan, Khan 1987, Rodriguez 1982), Pareto, Lomax distributions ($\beta = 1$) (see Lin 1988b), log–logistic distributions ($\beta > 0$, $p = 0$, $q = -1$, $d = \beta$, $c = -1/\beta$) (see Balakrishnan, Malik, Ahmed 1988);

iii) $p \neq -1$, $q = p+1$:

$$h(t) = \frac{1}{dq} t^q + c,$$

$\beta = 0$, $c \in \mathbb{R}$:

$$F(x) = (dq(\log x - c))^{1/q}, \quad \begin{cases} x \in (e^c, e^{c+1/(dq)}), & q > 0 \\ x \in (0, e^{c+1/(dq)}), & q < 0 \end{cases};$$

i.e. log power function distributions ($q > 0$), reflected log Pareto distributions ($q < 0$);

$\beta = 1 \vee (\beta > 0 \wedge c \geq 0)$, $q > 0$:

$$F(x) = (dq(\tfrac{1}{\beta} x^\beta - c))^{1/q}, \quad x \in ((\beta c)^{1/\beta}, (\beta(c + \tfrac{1}{dq}))^{1/\beta});$$

i.e. power function distributions ($\beta = q = 1$ in Lin 1988b);

$\beta = 1$, $c \in \mathbb{R}$, $q < 0$:

$$F(x) = (dq(x - c))^{1/q}, \quad x \in (-\infty, c + \tfrac{1}{dq});$$

i.e. reflected Pareto distributions;

$\beta = 2k - 1$, $k \in \mathbb{N}$, $c \in \mathbb{R}$, $q < 0$:

$$F(x) = (dq(\tfrac{1}{\beta} x^\beta - c))^{1/q}, \quad x \in (-\infty, (\beta(c + \tfrac{1}{dq}))^{1/\beta});$$

i.e. reflected Burr XII distributions;

iv) $p = -1$, $q = 0$:

$$h(t) = \frac{1}{d} \log t + c,$$

$\beta = 0$, $c \in \mathbb{R}$:

$$F(x) = e^{-cd} x^d, \quad x \in (0, e^c);$$

i.e. power function distributions (Malik 1967);

$\beta = 1$, $c \in \mathbb{R}$:

$$F(x) = \exp\{d(x-c)\}, \quad x \in (-\infty, c);$$

i.e. reflected exponential distributions;

$\beta = 2k-1$, $k \in \mathbb{N}$, $c \in \mathbb{R}$:

$$F(x) = \exp\{d(\tfrac{1}{\beta} x^{\beta} - c)\}, \quad x \in (-\infty, (\beta c)^{1/\beta};$$

i.e. reflected Weibull distributions;

v) $p = -1$, $q = -1$:

$$h(t) = \tfrac{1}{d} \log \tfrac{t}{1-t} + c,$$

$\beta = 0$, $c \in \mathbb{R}$:

$$F(x) = (1 + e^{cd} x^{-d})^{-1}, \quad x \in (0, \infty);$$

i.e. transformed Burr III distributions;

$\beta = 1$, $c \in \mathbb{R}$:

$$F(x) = (1 + \exp\{-d(x-c)\})^{-1}, \quad x \in (-\infty, \infty);$$

i.e. logistic distributions (Shah 1970, Khan, Yaqub, Parvez 1983, Balakrishnan, Malik, Ahmed 1988, Lin 1988b).

In the case of o OS's we obtain the relation (see Theorem 2.2.1.)

$$EX^{\alpha+\beta}_{r_0, n_0} - EX^{\alpha+\beta}_{r_0-1, n_0} = (\alpha + \beta)\, C\, EX^{\alpha}_{r_0+p, n_0+q}$$

(r_0, n_0 fixed) which is valid for correspondingly parametrized distributions with respect to the parameters β, p and q.

In Example 2.2.2. some distributions are shown possessing such an identity and the following diagram illustrates their places within the class of distributions considered.

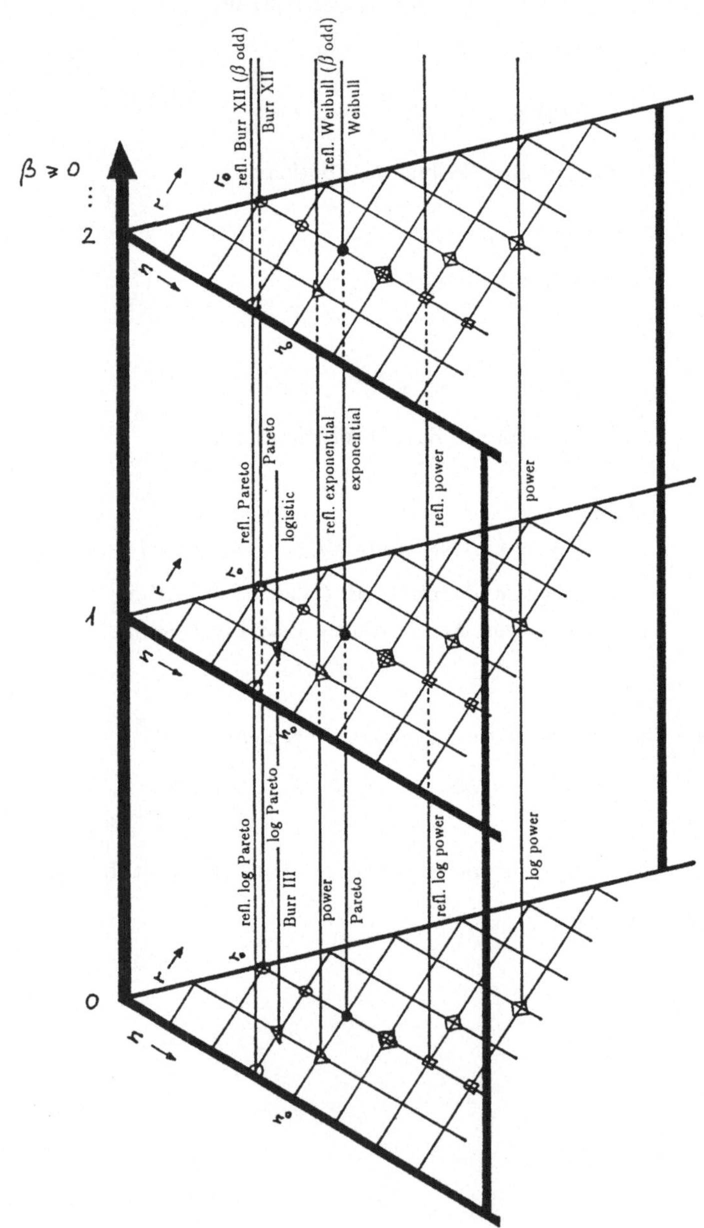

REMARKS 2.2.3. In the case $\beta = 0$, we find identities involving three moments of the same order. In particular, two parameter combinations lead to well known relations given by Malik (1966, 1967) for two moments of o OS's from Pareto and power function distributions :

i) $p = 0$, $q = 0$: The identity $\qquad EX^{\alpha}_{r,n} = \frac{n-r+1}{n-r+1-\alpha/d} EX^{\alpha}_{r-1,n}$

is valid for Pareto distributions with $F(x) = 1 - e^{cd} x^{-d}$, $x \in (e^c,\infty)$, $c \in \mathbb{R}$, $d > 0$, and it is stated in Malik (1966).

ii) $p = -1$, $q = 0$: The identity $\qquad EX^{\alpha}_{r,n} = (1+\frac{\alpha}{d(r-1)}) EX^{\alpha}_{r-1,n}$

is valid for power function distributions with $F(x) = e^{-cd} x^d$, $x \in (0,e^c)$, $c \in \mathbb{R}$, $d > 0$, and it is stated in Malik (1967).

EXAMPLES 2.2.4. in the case of record values ($m = -1$):

i) $p = 0$, $s \neq 0$:

$$h(t) = -\tfrac{1}{ds}(1-t)^s + c,$$

i.e. Example 2.2.2.ii) with q replaced by s;

ii) $p \neq -1$, $s = 0$:

$$h(t) = \frac{1}{d(p+1)} (\log \frac{1}{1-t})^{p+1} + c,$$

$\beta = 0$, $c \in \mathbb{R}$:

$$F(x) = 1 - \exp\{-(d(p+1)(\log x - c))^{1/(p+1)}\}, \quad \begin{cases} x \in (e^c,\infty), & p \geq 0 \\ x \in (0,e^c), & p < -1 \end{cases};$$

e.g. Pareto distributions ($p = 0$) (cf. Example 2.2.2.i));

$\beta = 1 \vee (\beta > 0 \wedge c \geq 0)$, $p \geq 0$:

$$F(x) = 1 - \exp\{-(d(p+1)(\tfrac{1}{\beta} x^{\beta} - c))^{1/(p+1)}\}, \quad x \in ((\beta c)^{1/\beta},\infty);$$

i.e. Weibull distributions and exponential distributions, if $\beta = 1$, $p = 0$, or $c = 0$, $p = \beta - 1$;

$\beta = 1$, $c \in \mathbb{R}$, $p < -1$:

$$F(x) = 1 - \exp\{-(d(p+1)(x-c))^{1/(p+1)}\}, \quad x \in (-\infty,c).$$

2.3. The Class $\mathcal{F}_C$ of Distributions

We have shown how to obtain parametrized recurrence relations for moments of o OS's and g OS's with respect to corresponding and, in this sense, related distributions. We give a third example considering a class of distributions that includes Cauchy and doubly truncated Cauchy distributions. First, we state the recurrence relation and the corresponding characterization result in the case of o OS's (see Kamps 1990a); then we consider g OS's.

Concerning the Cauchy distribution with density function

$$f(x) = \frac{1}{\pi}\frac{1}{1+x^2}, \quad -\infty < x < \infty,$$

characterization theorems based on identical distributions are cited in Johnson, Kotz (1970) and two theorems in connection with stable distributions are shown in Kagan, Linnik, Rao (1973, p 447/8). Here, a recurrence relation for moments of o OS's is used (see Barnett 1966) as a characterizing property.

A Cauchy distributed random variable X does not possess a finite expectation; however, Sen's theorem (cf. II.1.2.1. and II.1.2.2.) ensures the existence of $E|X_{r,n+1}^{\alpha+1}|$ for all $1 \leq r \leq n+1$ and n, $\alpha \in \mathbb{N}_0$ with $\alpha + 2 \leq r \leq n - \alpha$.

In this case, Barnett (1966) obtains a representation for the moments of OS's and by this he derives the following identity ($EX_{r,n}^0 = 0$) :

$$EX_{r,n}^{\alpha} - EX_{r-1,n}^{\alpha} = \alpha\frac{\pi}{n+1}(EX_{r,n+1}^{\alpha+1} - EX_{r,n+1}^{\alpha-1}), \quad \alpha \in \mathbb{N}_0 .$$

For moments of OS's from doubly truncated Cauchy distributions with density functions

$$f_t(x) = \frac{1}{\arctan b - \arctan a}\frac{1}{1+x^2}, \quad a < x < b,$$

Khan, Yaqub, Parvez (1983) derive :

$$EX_{r,n}^{\alpha} - EX_{r-1,n-1}^{\alpha} = \alpha\,(\arctan b - \arctan a)\frac{n-r+1}{n(n+1)}(EX_{r,n+1}^{\alpha+1} - EX_{r,n+1}^{\alpha-1}) .$$

Putting $a = -\infty$, $b = \infty$ and using Corollary 1.1. we have Barnett's identity.

2.3.1. Let $\mathcal{F}_C$ be the class of distribution functions F defined by

$$F_{1,q}(t) = \left(\frac{\arctan t - \arctan a}{\arctan b - \arctan a}\right)^{1/q} \text{ or } F_{2,q}(t) = 1 - \left(\frac{\arctan b - \arctan t}{\arctan b - \arctan a}\right)^{1/q}$$

with $t \in (a,b)$ and $q \in \mathbb{N}$.

THEOREM 2.3.2. Let the OS's be based on the distribution function $F \in \mathcal{F}_C$ with density function f, let $\alpha \geq 1$ be a constant and $F^{-1}(0) \geq 0$, if $\alpha \notin \mathbb{N}$.

Moreover, let
$$p = \begin{cases} q-1, & \text{if } F = F_{1,q} \\ 0, & \text{if } F = F_{2,q} \end{cases}.$$

Then for all $r, n \in \mathbb{N}$ satisfying $2 \leq r \leq n$ and

$$-\infty < EX^{\alpha}_{r,n},\ EX^{\alpha}_{r-1,n},\ EX^{\alpha+1}_{r+p,n+q} < \infty,$$

we have

$$EX^{\alpha}_{r,n} - EX^{\alpha}_{r-1,n} = \alpha\, q\, (\arctan b - \arctan a) \frac{\binom{n}{r-1}}{(r+p)\binom{n+q}{r+p}} \cdot \left(EX^{\alpha+1}_{r+p,n+q} + EX^{\alpha-1}_{r+p,n+q}\right).$$

PROOF Let $F \in \mathcal{F}_C$.

Since $F^{p}(t)\,(1-F(t))^{q-p-1} f(t) = \frac{1}{q} \frac{1}{\arctan b - \arctan a} \frac{1}{1+t^2}$

and using a formula in Khan, Yaqub, Parvez (1983) (see Theorem II.1.3.4. for g OS's), we obtain

$$\begin{aligned}
&EX^{\alpha+1}_{r+p,n+q} + EX^{\alpha-1}_{r+p,n+q} \\
&= (r+p)\binom{n+q}{r+p} \int_{-\infty}^{\infty} t^{\alpha-1} F^{r+p-1}(t)\,(1-F(t))^{n+q-r-p}\,(1+t^2)\, f(t)\, dt \\
&= \frac{1}{q} \frac{1}{\arctan b - \arctan a} (r+p)\binom{n+q}{r+p} \int_{-\infty}^{\infty} t^{\alpha-1} F^{r-1}(t)\,(1-F(t))^{n-r+1}\, dt \\
&= \frac{1}{q} \frac{1}{\arctan b - \arctan a} \frac{(r+p)\binom{n+q}{r+p}}{\alpha\binom{n}{r-1}} \left(EX^{\alpha}_{r,n} - EX^{\alpha}_{r-1,n}\right)
\end{aligned}$$

and thus the assertion.

REMARK If we assume (by analogy with Section 2.1.) that a recurrence relation of the above type (with parameters a,b,p,q) is valid for a sequence of pairs of indices $((r_j,n_j))_{j\in\mathbb{N}}$ such that the sequence $\{(1-t)^{n_j}\}_{j\in\mathbb{N}}$ or $\{t^{r_j-1}(1-t)^{n_j-r_j}\}_{j\in\mathbb{N}}$ of polynomials is complete on $L(0,1)$ (cf. Section II.2.1.), then the corresponding distribution function in $\mathcal{F}_C$ is characterized.

THEOREM 2.3.3. Let the OS's be based on some distribution function F, F^{-1} absolutely continuous in (0,1), let $a = F^{-1}(0)$, $b = F^{-1}(1)$ and let $\alpha \geq 1$ be a constant with

$$\begin{cases} |\{t \in (0,1); F^{-1}(t) = 0\}| \in \{0,1\}, & \alpha \in \mathbb{N}^{\geq 2} \\ F^{-1}(t) > 0 \quad \text{for all} \quad t \in (0,1), & \alpha \notin \mathbb{N} \end{cases}$$

Moreover, let $q \in \mathbb{N}$, $p \in \{0, q-1\}$, let $((r_j,n_j))_{j\in\mathbb{N}}$ be a sequence as described above, let the corresponding integrability conditions be fulfilled and let

$$-\infty < EX^{\alpha}_{r_j,n_j}, EX^{\alpha}_{r_j-1,n_j}, EX^{\alpha+1}_{r_j+p,n_j+q} < \infty$$

for all $j \in \mathbb{N}$.
If the identity

$$EX^{\alpha}_{r,n} - EX^{\alpha}_{r-1,n} = \alpha q(\arctan b - \arctan a)\frac{\binom{n}{r-1}}{(r+p)\binom{n+q}{r+p}} \cdot (EX^{\alpha+1}_{r+p,n+q} + EX^{\alpha-1}_{r+p,n+q})$$

is valid for all $(r,n) \in ((r_j,n_j))_{j\in\mathbb{N}}$, then F is given by

$$F(t) = \begin{cases} F_{1,q}(t), & \text{if } p = q-1 \\ F_{2,q}(t), & \text{if } p = 0 \end{cases}, \quad t \in (a,b).$$

PROOF Applying a representation for the difference of moments of successive OS's (see Lin 1988b, see Theorem II.1.3.1. for g OS's) via the pseudo–inverse, we obtain :

$$(F^{-1})'(t) = q(\arctan b - \arctan a)\, t^p (1-t)^{q-p-1} (1 + (F^{-1}(t))^2) \text{ a.e. on } (0,1).$$

By analogy with Theorem 2.1.6., we obtain the corresponding version for g OS's; the class $\mathcal{F}_C$ of distributions is then enlarged with respect to the new parameter m. (The choice m = −1 is excluded.).

REMARK 2.3.4.

For $p \geq 0$ and $s + (q-p)(m+1) = 1$, let $F_{1,q}^{(m)}$ be determined by

$$(m+1)^{-p} \sum_{j=0}^{p} (-1)^j \binom{p}{j} g_{j(m+1)}(F_{1,q}^{(m)}(x)) = p! \frac{\arctan x - \arctan a}{\arctan b - \arctan a} \left(\prod_{j=1}^{p} (j(m+1)+1) \right)^{-1} ;$$

for $p = 0$ and $s + q(m+1) > 0$, let $F_{2,q}^{(m)}$ be determined by

$$F_{2,q}^{(m)}(x) = 1 - \left(\frac{\arctan b - \arctan x}{\arctan b - \arctan a} \right)^{1/(s+q(m+1))}$$

If the appearing moments exist, we find :

$EX^{\alpha}(r,n,m,k) - EX^{\alpha}(r-1,n,m,k)$

$$= \alpha \frac{c_{r-2}(n,k)}{c_{r+p-1}(n+q,k+s)} \frac{(r+p-1)!}{(r-1)!} \frac{1}{D} (EX^{\alpha+1}(r+p,n+q,m,k+s) + EX^{\alpha-1}(r+p,n+q,m,k+s))$$

$$\text{with } D = \frac{1}{\arctan b - \arctan a} \begin{cases} p! \left(\prod_{j=1}^{p} (j(m+1)+1) \right)^{-1}, & p \geq 0 \text{ and } s+(q-p)(m+1)=1 \\ (s + q(m+1))^{-1}, & p = 0 \text{ and } s+q(m+1)>0 \end{cases}$$

2.4. Further Examples for Specific Distributions

In this section, some special cases of Theorem 2.2.1. and some other recurrence relations for power function, Pareto and Weibull distributions are shown.
Power function distributions appear in 2.2. as examples to Theorem 2.2.1. (especially in the case of o OS's). The following corollary shows a more general version with respect to the parameter m (see also Corollary 1.1.).

COROLLARY 2.4.1. If we have g OS's from a power function distribution with

$$F(x) = d\left(\tfrac{1}{\beta} x^{\beta} - c\right), \quad x \in ((c\beta)^{1/\beta}, (\beta(c+\tfrac{1}{d}))^{1/\beta}), \quad c \in \mathbb{R}, d > 0 \text{ and } \beta > 0,$$

then we find the following identities for $\alpha \geq 0$ and $q(m+1) + s = 1$
$(q \in \mathbb{Z}, s \in \mathbb{R}, k + s \geq 1)$:

$$EX^{\alpha+\beta}(r,n,m,k) - EX^{\alpha+\beta}(r-1,n,m,k)$$

$$\left(\overset{m\neq -1}{=} \frac{k+(n-1)(m+1)}{(r-1)(m+1)} \left(EX^{\alpha+\beta}(r-1,n-1,m,k) - EX^{\alpha+\beta}(r-1,n,m,k)\right)\right)$$

$$= \frac{k+(n-1)(m+1)}{k+(n-r)(m+1)} \left(EX^{\alpha+\beta}(r,n,m,k) - EX^{\alpha+\beta}(r-1,n-1,m,k)\right)$$

$$= \frac{\alpha+\beta}{d} \frac{c_{r-2}(n,k)}{c_{r-1}(n+q,k+s)} EX^{\alpha}(r,n+q,m,k+s)\,, \quad r \geq 2\,;$$

if $q = 0$ and $s = 1$ we have for an arbitrary m :

$$EX^{\alpha+\beta}(r,n,m,k) - EX^{\alpha+\beta}(r-1,n,m,k) = \frac{\alpha+\beta}{d} \frac{c_{r-2}(k)}{c_{r-1}(k+1)} EX^{\alpha}(r,n,m,k+1)\,, \quad r \geq 2\,.$$

PROOF directly follows from Corollary 1.1. and Theorem 2.2.1. with $\beta > 0$, $p = 0$ and $q(m+1) + s - 1 = 0$; hence $h'(t) = \frac{1}{d}$ a.e. on $(0,1)$.

REMARKS

i) Concerning moments of g OS's of order α from a power function distribution with $F(x) = c\,x^{a}\,,\ x \in (0\,,\,c^{-1/a})\,,\ a,c > 0\,,$ the explicit expression

$$EX^{\alpha}(r,n,m,k) = c_{r-1}(k)\, c^{-\alpha/a} \sum_{j=0}^{\alpha/a} (-1)^{j} \binom{\alpha/a}{j} \frac{1}{c_{r-1}(k+j)}$$

yields the assertion of Corollary 2.4.1. for $\alpha/a \in \mathbb{N}$, only.

ii) Lin (1988b) states the identity for o OS's from a uniform distribution ($\beta = 1$, $m = 0$, $k = 1$, $s = 0$, $q = 1$).

iii) Putting $\alpha = 0$, we obtain an explicit expression for the difference $EX^{\beta}(r,n,m,k) - EX^{\beta}(r-1,n,m,k)$, $\beta > 0$.

If $k + (n-r)(m+1) > \alpha/a$ and $k + (n-1)(m+1) > \alpha/a$ we have

$$EX^{\alpha}(r,n,m,k) = c^{\alpha/a} \frac{c_{r-1}(k)}{c_{r-1}(k - \alpha/a)}$$

for g OS's from an underlying Pareto distribution (see II.1.4.2.) with distribution function

$$F(x) = 1 - c/x^a \,, \quad a,c > 0 \,, \; x \in (c^{1/a}, \infty) \,.$$

This representation leads to special recurrence relations which contain some known results for o OS's.

Lemma 2.4.2. If g OS's are based on a Pareto distribution with

$$F(x) = 1 - c/x^a \,, \quad x \in (c^{1/a}, \infty) \,, \; a,c > 0 \,,$$

then we observe :

i) $$EX^{\alpha}(r,n,m,k) = c^{1/a} \frac{c_{r-1}(k - \frac{\alpha-1}{a})}{c_{r-1}(k - \alpha/a)} EX^{\alpha-1}(r,n,m,k) \,,$$

ii) $$EX^{\alpha}(r,n,m,k) = \frac{c_{r-1}(k)}{c_{r-1}(k')} \frac{c_{r-1}(k' - \alpha/a)}{c_{r-1}(k - \alpha/a)} EX^{\alpha}(r,n,m,k') \,,$$

iii) $$EX^{\alpha}(r,n,m,k) = \frac{k + (n-r)(m+1)}{k - \alpha/a + (n-r)(m+1)} EX^{\alpha}(r-1,n,m,k) \,, \; r \geq 2 \,,$$

iv) $$EX^{\alpha}(r,n,m,k) = \frac{k + (n-1)(m+1)}{k + (n-r-1)(m+1)} \frac{k - \alpha/a + (n-r-1)(m+1)}{k - \alpha/a + (n-1)(m+1)} EX^{\alpha}(r,n-1,m,k) \,,$$

$r \leq n-1$,

v) $$EX^{\alpha}(r,n,m,k) = \frac{k + (n-1)(m+1)}{k - \alpha/a + (n-1)(m+1)} EX^{\alpha}(r-1,n-1,m,k) \,, \; r \geq 2 \,.$$

Remark In the case of o OS's ($m = 0$, $k = 1$), iii) can be found in Malik (1966) and v) in Balakrishnan, Joshi (1982).

Corollary 2.4.3. If we have g OS's from a Pareto distribution with

$$F(x) = 1 - \left(-\frac{d}{\beta}(q(m+1) + s)\, x^{\beta}\right)^{\frac{1}{q(m+1)+s}} , \quad x \in \left[\left(-\frac{\beta}{d}\frac{1}{q(m+1)+s}\right)^{1/\beta}, \infty\right) ,$$

$d > 0$, $\beta > 0$, then we find the following identities for $q(m+1) + s < 0$ ($q \in \mathbb{Z}$, $s \in \mathbb{R}$, $k + s \geq 1$):

$$EX^{\alpha+\beta}(r,n,m,k) - EX^{\alpha+\beta}(r-1,n,m,k)$$

$$\left(\overset{m\neq-1}{=} \frac{k+(n-1)(m+1)}{(r-1)(m+1)} (EX^{\alpha+\beta}(r-1,n-1,m,k) - EX^{\alpha+\beta}(r-1,n,m,k))\right)$$

$$= \frac{k+(n-1)(m+1)}{k+(n-r)(m+1)} (EX^{\alpha+\beta}(r,n,m,k) - EX^{\alpha+\beta}(r-1,n-1,m,k))$$

$$= \frac{\alpha+\beta}{d} \frac{c_{r-2}(n,k)}{c_{r-1}(n+q,k+s)} EX^{\alpha}(r,n+q,m,k+s)\,, \quad r \geq 2\,;$$

if $q = 0$ and $s < 0$ $(k + s \geq 1)$, we have for an arbitrary m :

$$EX^{\alpha+\beta}(r,n,m,k) - EX^{\alpha+\beta}(r-1,n,m,k) = \frac{\alpha+\beta}{d} \frac{c_{r-2}(k)}{c_{r-1}(k+s)} EX^{\alpha}(r,n,m,k+s)\,, \quad r \geq 2$$

(subject to the existence of the appearing moments).

Proof directly follows from Corollary 1.1. and Theorem 2.2.1. with $\beta > 0$, $p = 0$, $c = 0$ and $q(m+1) + s < 0$; hence, $h'(t) = \frac{1}{d}(1-t)^{q(m+1)+s-1}$ a.e. on $(0,1)$.

Remarks

i) Lin (1988b) states the identity for o OS's ($m = 0$, $k = 1$, $q < 0$, $s = 0$, $\beta = 1$).

ii) Putting $\alpha = 0$, we obtain an explicit expression for the difference $EX^{\beta}(r,n,m,k) - EX^{\beta}(r-1,n,m,k)$, $\beta > 0$.

Let $m = -1$; then we have

$$EX^{\alpha}(r,n,-1,k) = \frac{1}{(r-1)!}(c\,k)^{-\alpha/a}\,\Gamma(\alpha/a + r)\,, \text{ if } \alpha/a + 1 > 0\,,$$

for records from an underlying Weibull distribution (see II.1.4.3.) with distribution function

$$F(x) = 1 - \exp\{-c\,x^a\}\,, \quad x \in (0,\infty)\,,\ a,c > 0\,.$$

This representation leads to some special recurrence relations.

LEMMA 2.4.4. If a Weibull distribution with

$$F(x) = 1 - \exp\{-c\, x^a\}\,, \quad x \in (0,\infty)\,,\ a,c > 0\,,$$

is the underlying distribution of g OS's with $m = -1$ (i.e. k–th records), then we have the following recursions :

i) $$EX^\alpha(r,n,-1,k) = (c\,k)^{-1/a} \frac{\Gamma(\alpha/a + r)}{\Gamma(\frac{\alpha-1}{a} + r)} EX^{\alpha-1}(r,n,-1,k)\,,$$

ii) $$EX^\alpha(r,n,-1,k) = (k/k')^{-\alpha/a} EX^\alpha(r,n,-1,k')\,,$$

iii) $$EX^\alpha(r,n,-1,k) = (\frac{\alpha}{a(r-1)} + 1)\ EX^\alpha(r-1,n,-1,k)\,,\ r \geq 2\,,$$

iv) $$EX^\alpha(r,n,-1,k) - EX^\alpha(r-1,n,-1,k) = \frac{\alpha}{a}(c\,k)^{-1/a}\frac{\Gamma(\alpha/a + r - 1)}{\Gamma(\frac{\alpha-1}{a}+r)} EX^{\alpha-1}(r,n,-1,k)$$

$$\left(= \frac{\alpha}{c\,k} EX^{\alpha-1}(r,n,-1,k) \text{ for exponential distributions, } a = 1\right)$$

with $EX^\alpha(0,n,-1,k) = 0$.

REMARKS We refer to Theorem 2.1.4. in the case $a = 1$, i.e. exponential distributions. Identity iv) is a special case of Theorem 2.2.1. (see also Lin 1988b).

In this section we show how to proceed to obtain recurrence relations for moments of g OS's. Other possibilities are shown in the literature in the case of o OS's.
For o OS's and records, Too, Lin (1989) give identities involving moments of first and second orders which lead to characterizations of power function and Weibull distributions without any further assumptions. The corresponding result for g OS's is shown in Section 3.

David (1981), e.g., considers bounds for moments of o OS's. If it is possible to characterize equality in such an inequality, we obtain characterizations of probability distributions on the one hand and recurrence relations for certain distributions on the other. Lin (1988a) applies the Cauchy–Schwarz inequality to characterize the uniform distribution. In Gajek, Gather (1989,1991) and Kamps (1991a), this result is generalized and other integral inequalities are applied to obtain characterizations and related recurrence relations for Pareto, power function, Burr, Fisher and Weibull distributions. Inequalities for moments of g OS's are examined in the following chapter.

3. Characterizations of Distributions by Two Moments

Too, Lin (1989) show characterizing recurrence relations for moments of o OS's and records from special power function and Weibull distributions, respectively. If the existence of the appearing moments is ensured, then we have for $p \in \mathbb{N}$:

i) $$(r\tbinom{n}{r})^{-1} EX_{r,n}^2 - 2((r+p)\tbinom{n+p}{r+p})^{-1} EX_{r+p,n+p} + ((r+2p)\tbinom{n+2p}{r+2p})^{-1} = 0,$$

iff the underlying distribution function F is given by

$$F(x) = x^{1/p}, \quad x \in (0,1);$$

ii) $$(r-1)!\ EX_{L(r)}^2 - 2(r-1+p)!\ EX_{L(r+p)} + (r-1+2p)! = 0,$$

iff the underlying distribution function F is given by

$$F(x) = 1 - \exp\{-x^{1/p}\}, \quad x \in (0,\infty).$$

Recursions of such type are also available for g OS's and, by this, characterizations of Burr XII and Weibull distributions result.

THEOREM 3.1. Let $p \in \mathbb{N}$ and $X(r,n,m,k)$, $X(r+p,n+p,m,k)$ be g OS's based on F with finite second and first moment respectively, such that $k + (n + 2p - 1)(m + 1) > 0$. Then we find :

$$\frac{(r-1)!}{c_{r-1}(n)} EX^2(r,n,m,k) - 2\frac{(r+p-1)!}{c_{r+p-1}(n+p)} EX(r+p,n+p,m,k) + \frac{(r+2p-1)!}{c_{r+2p-1}(n+2p)} = 0,$$

iff the underlying distribution function F is given by

$$F(x) = g_m^{-1}(x^{1/p}) = \begin{cases} 1 - (1 - (m+1)\, x^{1/p})^{1/(m+1)}, & \begin{cases} x \in (0,1), & m > -1 \\ x \in (0,\infty), & m < -1 \end{cases} \\ 1 - \exp\{-x^{1/p}\}, & x \in (0,\infty), \quad m = -1 \end{cases}.$$

PROOF Using II.1.1.1. and

$$\frac{(r-1)!}{c_{r-1}(n)} EX^2(r,n,m,k) - 2\frac{(r+p-1)!}{c_{r+p-1}(n+p)} EX(r+p,n+p,m,k) + \frac{(r+2p-1)!}{c_{r+2p-1}(n+2p)}$$

$$= \int_0^1 (F^{-1}(t) - g_m^p(t))^2 \, (1-t)^{k+(n-r)(m+1)-1} \, g_m^{r-1}(t) \, dt \, ,$$

the assertion follows noticing

$$g_m^{-1}(x) = \begin{cases} 1-(1-(m+1)\,x)^{1/(m+1)} & , \ m \neq -1 \\ 1-e^{-x} & , \ m = -1 \end{cases} .$$

Thus, the results of Too, Lin (1989) for o OS's and records are contained in a result valid for g OS's. Fixing m, each g OS $X(r,n,m,k)$ satisfies a corresponding recurrence relation which is parametrized with respect to r, n and k.

By analogy with Too, Lin (1989), we refer to the case $r = n = 1$, $p = 1$ which leads to the equation

$$EX^2(1,1,m,k) - \frac{2}{k+m+1} EX(2,2,m,k) + \frac{2}{(k+m+1)(k+2m+2)} = 0 .$$

The distributions which are characterized by this identity are determined by

$$F_m(x) = \begin{cases} 1-(1-(m+1)\,x)^{\frac{1}{m+1}} & , \ m \neq -1 \\ 1-e^{-x} & , \ m = -1 \end{cases} .$$

Hence, the second moment of $X(1,1,m,k)$ is given by

$$EX^2(1,1,m,k) = k \int_0^1 (F_m^{-1}(t))^2 \, (1-t)^{k-1} \, dt = \frac{2}{(k+m+1)(k+2m+2)} ,$$

if $k + 2m + 2 > 0$ for $m < -1$ is assumed.

Using $EX(2,2,m,k) = \frac{2}{k+2m+2}$, we find :

Corollary 3.2. Let $k + 2m + 2 > 0$. Then the moment equations

$$EX^2(1,1,m,k) = \frac{2}{(k+m+1)(k+2m+2)} \quad \text{and} \quad EX(2,2,m,k) = \frac{2}{k+2m+2}$$

are valid iff the underlying distribution has the distribution function F_m.

In particular, the results of Too, Lin (1989) are contained :

$$EX^2 = \frac{1}{3} \text{ and } EX_{2,2} = \frac{2}{3} \Leftrightarrow F(x) = x\,,\; x \in (0,1)\,;$$

$$F \text{ continuous, } EX^2 = EX_{L(2)} = 2 \Leftrightarrow F(x) = 1 - e^{-x}\,,\; x \in (0,\infty)\,.$$

The results are already stated in Lin (1988a) as corollaries of assertions for moments of o OS's and records derived by the Cauchy inequality (see Theorem IV.1.1.). Motivated by this, Lin (1988a) poses two problems :

i) Does some $n \geq 3$ exist, such that $EX^2 = \frac{1}{3}$ and $EX_{n,n} = \frac{n}{n+1}$ characterize the uniform distribution ?

ii) Does some $n \geq 3$ exist, such that (for F being continuous) $EX^2 = 2$ and $EX_{L(n)} = n$ characterize the exponential distribution?

The answer to the questions is 'no', since Too, Lin (1989) give counterexamples for any $n \geq 3$ $(a > 0)$:

ad i) $F(x) = (\frac{x+b}{a})^{1/(n-1)}\,,\; x \in (-b,a-b)\,,$

ad ii) $F(x) = 1 - \exp\{-(\frac{x+b}{a})^{1/(n-1)}\}\,,\; x \in (-b,\infty)\,.$

Chapter IV

Inequalities for Moments of Generalized Order Statistics and Characterizations of Distributions

Bounds and approximations for moments of o OS′s are often considered in the literature. Detailed surveys on related results can be found in David (1981, Ch.4) and Arnold, Balakrishnan (1989).

First results on bounds for moments of o OS′s date back to Plackett (1947) and Moriguti (1951) and are then generalized by Gumbel (1954) and Hartley, David (1954). We have :

Let $X_1,\ldots,X_n$ be iid random variables with distribution function F and $EX_1 = 0$, $EX_1^2 = 1$.

Then
$$EX_{n,n} \le \frac{n-1}{(2n-1)^{1/2}},$$

and we find equality iff F is a special power function distribution :

$$F(x) = \left(\frac{1+bx}{n}\right)^{1/(n-1)}, \; b = \frac{n-1}{(2n-1)^{1/2}}, \quad x \in \left(-\frac{(2n-1)^{1/2}}{n-1}, (2n-1)^{1/2}\right).$$

Nagaraja (1978) states the following result for records :

Let $(X_i)_{i\in\mathbb{N}}$ be a sequence of iid random variables with continuous distribution function F and $EX_1 = 0$, $EX_1^2 = 1$.

Then
$$EX_{L(n)} \le \left(\binom{2n}{n} - 1\right)^{1/2}$$

and equality characterizes special Weibull distributions.

Many other results are listed in Arnold, Balakrishnan (1989, Ch.6). For further details we also refer to, e.g., Terrell (1983), Arnold (1985, 1988), David (1988), Papathanasiou (1990), Balakrishnan (1993) and Balakrishnan, Balasubramanian (1993).

1. The Inequality of Hölder

Lin (1988a) applies the Cauchy–Schwarz inequality to moments of o OS's and record values to obtain characterization theorems for uniform and exponential distributions, respectively.

Theorem 1.1. (Lin 1988a) Let $X, X_1, X_2, \ldots$ be iid according to the distribution function F.

i) If $EX^2 < \infty$ and $2 \le r \le n$, then

$$(EX_{r,n})^2 \le \frac{rn}{(r-1)(n+1)} EX^2_{r-1,n-1}$$

with equality iff F is the distribution function of a degenerate distribution at 0 or of a uniform distribution on (0,c) for some $c > 0$.

ii) Let F be continuous, $E|X|^p < \infty$ for some $p > 2$, and $r \ge 2$. Then

$$(EX_{L(r)})^2 \le \frac{r}{r-1} EX^2_{L(r-1)}$$

with equality iff $F(x) = 1 - \exp\{-x/c\}$, $x > 0$, for some $c > 0$.

Thus, results of this type, i.e. characterizing equality in an inequality for moments, lead to bounds for certain moments of g OS's as well as to new recurrence relations for moments with respect to the distributions characterized by equality.

In particular, we observe that fixing only two or three moments is sufficient to determine the corresponding distributions uniquely; this fact is pointed out by Lin (1988a) (cf. Section III.3.).

Theorem 1.1. directly implies the example of Too, Lin (1989) given in III.3. This result is remarkable in view of the variety of characterization theorems (see Sections II.2.2. and III.2.1.) in which assumptions are imposed on a sequence of moments.

The results of Lin (1988a) are taken up in Gajek, Gather (1991) and Kamps (1991a) and are generalized with respect to appearing powers and indices of o OS's and records applying Hölder's inequality and its inverse version (see Mitrinović 1970, p 54, Beckenbach, Bellman 1961, p 21/2).

The structure of g OS's leads to a unified approach and the above cited results turn out to be special cases.

When comparing results for record values in the literature, one has to take note of the different numbering of records (cf. I.2.6.).

REMARK

Throughout in this chapter, the existence of all appearing moments is assumed. The existence may be ensured globally using results of Chapter II.

1.1. An Inequality Involving Three Moments

The following theorem gives bounds for single moments in terms of products of two moments of g OS's; in the case of o OS's and k–th record values this result is derived by Gajek, Gather (1989, 1991). The generalization to g OS's is now clear. Merely, we have to assume some regularity conditions to let the appearing g OS's be well defined.

THEOREM 1.1.1. (Gajek, Gather 1989 for $m \in \{0,-1\}$)

Let F be the underlying, non–degenerate distribution function of the appearing g OS's.

Let $\alpha_1,\alpha_2 \in \mathbb{R}$, $\alpha = \alpha_1 + \alpha_2$, $p_1,p_2 \in \mathbb{R}\backslash\{0,1\}$ with $\frac{1}{p_1} + \frac{1}{p_2} = 1$,

$r,r_1,r_2,\ n,n_1,n_2 \in \mathbb{N}$, $k,k_1,k_2 \geq 1$, $r \leq n$, $r_1 \leq n_1$, $r_2 \leq n_2$, $\tilde{m} = (m_1,\dots,m_{n-1})' \in \mathbb{R}^{n-1}$,

$\tilde{m}_1 = (m_1^{(1)},\dots,m_{n_1-1}^{(1)})' \in \mathbb{R}^{n_1-1}$, $\tilde{m}_2 = (m_1^{(2)},\dots,m_{n_2-1}^{(2)})' \in \mathbb{R}^{n_2-1}$,

satisfying the regularity conditions

$$m_1 = \dots = m_{r-1} = m_1^{(1)} = \dots = m_{r_1-1}^{(1)} = m_1^{(2)} = \dots = m_{r_2-1}^{(2)} = m\,,$$

$$k + n - i + M_i \geq 1\,,\ i = 1,\dots,n-1\,,$$

$$k_1 + n_1 - i_1 + M_{i_1}^{(1)} \geq 1\,,\ i_1 = 1,\dots,n_1-1 \quad \left(M_{i_1}^{(1)} = \sum_{j=i_1}^{n_1-1} m_j^{(1)} \right),$$

$$k_2 + n_2 - i_2 + M_{i_2}^{(2)} \geq 1 \, , \, i_2 = 1,\dots,n_2-1 \quad \left(M_{i_2}^{(2)} = \sum_{j=i_2}^{n_2-1} m_j^{(2)} \right)$$

and $F^{-1}(t) \neq 0$ for all $t \in (0,1)$, if $\alpha < 0 \vee \alpha_1 p_1 < 0 \vee \alpha_2 p_2 < 0$.

With respect to these quantities let the following assumption be valid :

> We have
>
> $$\frac{\left((k_1 + n_1 - r_1 + M_{r_1}^{(1)}) \right) - \left((k_2 + n_2 - r_2 + M_{r_2}^{(2)}) \right)}{\left((k + n - r + M_r) \right) - \left((k_2 + n_2 - r_2 + M_{r_2}^{(2)}) \right)} = \frac{r_1 - r_2}{r - r_2}$$
>
> and p_1 equals one of the two sides, if all nominators and denominators do not vanish. If one of them vanishes, then the corresponding nominator or denominator must also be zero and $p_1 \in \mathbb{R}\backslash\{0,1\}$ equals the well defined other ratio. If both ratios are of the form $0/0$, then $p_1 \in \mathbb{R}\backslash\{0,1\}$ is arbitrary.

Then we find for $p_1 > 1$ ($0 \neq p_1 < 1$) :

$$\frac{(r-1)!}{c_{r-1}(\tilde{m},k)} E|X(r,n,\tilde{m},k)|^{\alpha} \underset{(\geq)}{\leq} \left\{ \frac{(r_1 - 1)!}{c_{r_1-1}(\tilde{m}_1,k_1)} E|X(r_1,n_1,\tilde{m}_1,k_1)|^{\alpha_1 p_1} \right\}^{1/p_1}$$
$$\cdot \left\{ \frac{(r_2 - 1)!}{c_{r_2-1}(\tilde{m}_2,k_2)} E|X(r_2,n_2,\tilde{m}_2,k_2)|^{\alpha_2 p_2} \right\}^{1/p_2} .$$

Equality holds iff for some constant $c > 0$:

$$|F^{-1}(t)|^{\alpha_1 p_1 - \alpha_2 p_2} = c\,(1-t)^{(k_2+n_2-r_2+M_{r_2}^{(2)}) - (k_1+n_1-r_1+M_{r_1}^{(1)})} (g_m(t))^{r_2 - r_1}$$

a.e. on $(0,1)$.

Proof The assumptions yield

$$(k+n-r+M_r) - (k_2+n_2-r_2+M_{r_2}^{(2)}) = \frac{1}{p_1} \left((k_1+n_1-r_1+M_{r_1}^{(1)}) - (k_2+n_2-r_2+M_{r_2}^{(2)}) \right)$$

$$\Leftrightarrow\ k+n-r+M_r-1 = \frac{1}{p_1}(k_1+n_1-r_1+M^{(1)}_{r_1}-1) + \frac{1}{p_2}(k_2+n_2-r_2+M^{(2)}_{r_2}-1)$$

and $r-r_2 = \frac{1}{p_1}(r_1-r_2) \Leftrightarrow r-1 = \frac{1}{p_1}(r_1-1) + \frac{1}{p_2}(r_2-1)$.

Now we apply Hölder's inequality in the case of $p_1 > 1$ or its inverse version if $0 \neq p_1 < 1$ (cf. Mitrinović 1970, p 54, Beckenbach, Bellman 1961, p 21/2) to the moments $E|X(r,n,\tilde{m},k)|^{\alpha}$.

Let $p_1 > 1$:

$$\frac{(r-1)!}{c_{r-1}(\tilde{m},k)} E|X(r,n,\tilde{m},k)|^{\alpha} = \int_0^1 |F^{-1}(t)|^{\alpha} (1-t)^{k+n-r+M_r-1} g_m^{r-1}(t)\, dt$$

$$\leq \left(\int_0^1 |F^{-1}(t)|^{\alpha_1 p_1} (1-t)^{k_1+n_1-r_1+M^{(1)}_{r_1}-1} (g_m(t))^{r_1-1} dt\right)^{1/p_1}$$

$$\cdot \left(\int_0^1 |F^{-1}(t)|^{\alpha_2 p_2} (1-t)^{k_2+n_2-r_2+M^{(2)}_{r_2}-1} (g_m(t))^{r_2-1} dt\right)^{1/p_2}$$

$$= \left\{\frac{(r_1-1)!}{c_{r_1-1}(\tilde{m}_1,k_1)} E|X(r_1,n_1,\tilde{m}_1,k_1)|^{\alpha_1 p_1}\right\}^{1/p_1}$$

$$\cdot \left\{\frac{(r_2-1)!}{c_{r_2-1}(\tilde{m}_2,k_2)} E|X(r_2,n_2,\tilde{m}_2,k_2)|^{\alpha_2 p_2}\right\}^{1/p_2}.$$

In the case $0 \neq p_1 < 1$ we obtain the reversed inequality.

We obtain equality iff for some constant $c > 0$:

$$|F^{-1}(t)|^{\alpha_1 p_1 - \alpha_2 p_2} = c\,(1-t)^{(k_2+n_2-r_2+M^{(2)}_{r_2}) - (k_1+n_1-r_1+M^{(1)}_{r_1})} (g_m(t))^{r_2-r_1}$$

a.e. on $(0,1)$.

Remark 1.1.2. In the special case

$$m_1 = \dots = m_{n-1} = m^{(1)}_1 = \dots = m^{(1)}_{n_1-1} = m^{(2)}_1 = \dots = m^{(2)}_{n_2-1} = m,$$

containing well known results for o OS's $(m = 0)$ and k–th records $(m = -1)$, the ratio appearing in the assumption of the theorem simplifies to

$$\frac{k_1 - k_2 + (m+1)((n_1-r_1) - (n_2-r_2))}{k - k_2 + (m+1)((n-r) - (n_2-r_2))} = \frac{r_1 - r_2}{r - r_2}$$

(i.e. conditions (A) and ($\tilde{A}$) in Gajek, Gather 1989).

This theorem includes a variety of interesting special cases. In particular under the condition of 1.1.2., those cases may be noticed which lead to a vanishing exponent in the representation of the characterized distributions (via the pseudo–inverse). Gajek, Gather (1989) give many examples for $m = 0$ and $m = -1$. The results of Lin (1988a) are obtained putting $\alpha = \alpha_1 = 1$, $\alpha_2 = 0$, $p = 2$, $r_1 = r - 1$, $n_1 = n - 1$. When characterizing equality, the theorem gives recurrence relations and characterization results for Burr XII, power function, Pareto, exponential and Weibull distributions.

Remark 1.1.3.

The distributions characterized by equality in Theorem 1.1.1. are given by

$$|F^{-1}(t)|^{\alpha_1 p_1 - \alpha_2 p_2} = c\,(1-t)^{A_2 - A_1}\,(g_m(t))^{r_2 - r_1} \quad \text{a.e. on } (0,1)$$

with $A_i = k_i + n_i - r_i + M^{(i)}_{r_i}$, $i = 1,2$.

Concerning the exponents we consider the cases $A_1 = A_2$ and $r_1 = r_2$.

i) Let $A_1 = A_2$; i.e. $|F^{-1}(t)|^{\alpha_1 p_1 - \alpha_2 p_2} = c\,(g_m(t))^{r_2 - r_1}$ a.e. on $(0,1)$.

$A_1 = A_2$ and the assumption together imply $k + n - r + M_r = A_1$.

If $r_1 = r_2$ is excluded, then $p_1 = \dfrac{r_1 - r_2}{r - r_2}$.

If we assume $F^{-1}(t) > 0$ for all $t \in (0,1)$ for simplicity, we find $\dfrac{\tilde{\alpha}}{r_2 - r_1} > 0$ (because of monotonicity) (with $\tilde{\alpha} = \alpha_1 p_1 - \alpha_2 p_2$) and

$$F^{-1}(t) = (c\, g_m^{r_2 - r_1})^{1/\tilde{\alpha}} = \begin{cases} \left(c\,\left(\frac{1}{m+1}\,(1-(1-t)^{m+1}\right)^{r_2 - r_1}\right)^{1/\tilde{\alpha}}, & m \neq -1 \\ \left(c\,\left(\log\frac{1}{1-t}\right)^{r_2 - r_1}\right)^{1/\tilde{\alpha}}, & m = -1 \end{cases} \quad \text{a.e. on } (0,1).$$

Hence F is given by

$$F(x) = \begin{cases} 1 - \left(1 - (m+1)(\frac{x}{c})^{\tilde{\alpha}}\ ^{1/(r_2-r_1)}\right)^{1/(m+1)} & , m \neq -1 \\ 1 - \exp\left(- (\frac{x}{c})^{\alpha}\ ^{1/(r_2-r_1)} \right) & , m = -1 \end{cases}.$$

In this class of distributions, we observe power function distributions (m = 0), Burr XII distributions (m < −1) and Weibull distributions (m = −1) (exponential distribution if $\frac{\alpha_1 p_1 - \alpha_2 p_2}{r_2 - r_1} = 1$).

ii) Let $r_1 = r_2$; i.e. $|F^{-1}(t)|^{\tilde{\alpha}} = c\,(1-t)^{A_2-A_1}$ a.e. on (0,1).

$r_1 = r_2$ leads to $r = r_1 = r_2$.

Excluding a degenerate distribution ($A_1 = A_2$) and noticing

$A_1 - A_2 = (k_2+n_2+M^{(2)}_{r_2}) - (k_1+n_1+M^{(1)}_{r_1})$, we have

$$p_1 = \frac{\left((k_1 + n_1 + M^{(1)}_{r_1}) \right) - \left((k_2 + n_2 + M^{(2)}_{r_2}) \right)}{\left((k + n + M_r) \right) - \left((k_2 + n_2 + M^{(2)}_{r_2}) \right)}.$$

Assuming $F^{-1}(t) > 0$ for all $t \in (0,1)$, we find $\frac{\tilde{\alpha}}{A_2-A_1} > 0$ and

$$F^{-1}(t) = (c\,(1-t)^{A_2-A_1})^{1/\tilde{\alpha}} \quad \text{a.e. on } (0,1)\,.$$

Thus F is a Pareto distribution function with

$$F(x) = 1 - (\frac{x}{c})^{\tilde{\alpha}}\ ^{1/(A_2-A_1)}\ , \ x \geq c^{1/\tilde{\alpha}}\,.$$

EXAMPLE 1.1.4.

ad 1.1.3. i): In the case $c = 1$, $\frac{\tilde{\alpha}}{r_2-r_1} > 0$ and $\tilde{\alpha} = (r_2 - r_1)\,\beta$, $\beta = \frac{\alpha_2}{r_1-r} - \frac{\alpha_1}{r-r_2} > 0$, the characterized distributions are given by

$$F(x) = \begin{cases} 1-(1-(m+1)\,x^{\beta})^{1/(m+1)}, & \begin{cases} x \in (0,\,(m+1)^{-1/\beta}), & m > -1 \\ x \in (0,\,\infty), & m < -1 \end{cases} \\ 1-\exp(-\,x^{\beta}), & x \in (0,\,\infty), \quad m = -1 \end{cases}$$

with corresponding density functions

$$f(x) = \begin{cases} \beta\, x^{\beta-1}\,(1-(m+1)\,x^{\beta})^{-m/(m+1)}, & m \neq -1 \\ \beta\, x^{\beta-1}\,\exp(-\,x^{\beta}), & m = -1 \end{cases}.$$

Some graphs are shown in the following plots.

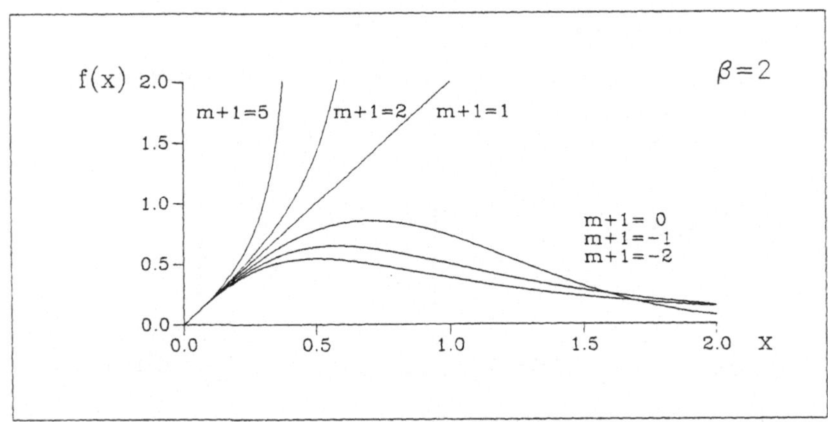

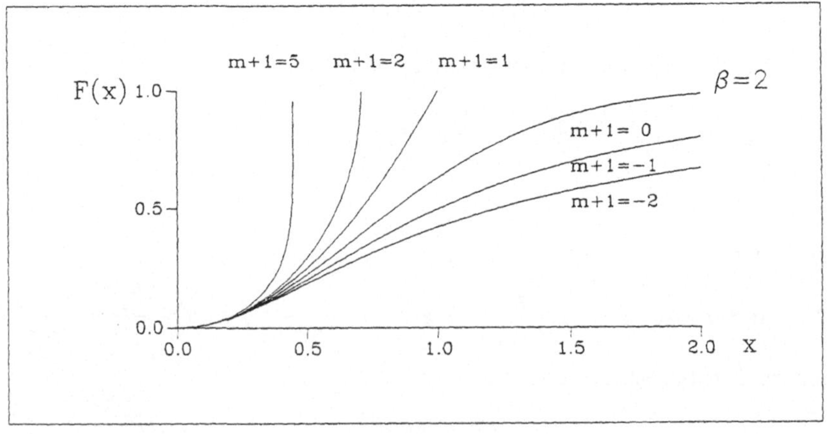

According to the assumption of Theorem 1.1.1., some first entries m_i, $m^{(1)}_{i_1}$, $m^{(2)}_{i_2}$ coincide concerning the three appearing types of g OS's, exept for the case $r = 1 \vee r_1 = 1 \vee r_2 = 1$. This condition is related to the representation of marginal densities (see I.3.1.2.) (uniqueness of the function g_m). Thus, in these cases we have inequalities at hand within a certain subclass of g OS's with respect to m. Choosing $r = 1 \vee r_1 = 1 \vee r_2 = 1$, we obtain inequalities involving, e.g., o OS's and records based on some distribution function F and characterization results for Pareto, power function and Weibull distributions. Putting $r_1 = r_2 = 1$ leads to $r = 1$ (because of the assumption of the theorem). However, in the case $r = 1 \vee r_1 = 1 \vee r_2 = 1$ we may apply transformation I.3.1.5. By this, the marginal density of the first record ($m = -1$) coincides with the density of the maximum of k random variables distributed according to the underlying distribution function. Hence, the corresponding moments of some minimum may be viewed as moments of some related first record and vice versa. The following corollaries are also stated in Gajek, Gather (1989) in a different notation.

COROLLARY 1.1.5. With the notations and under the assumptions of Theorem 1.1.1. we have :

i)
$$\frac{1}{n} E|X(1,n,0,1)|^{\alpha} \underset{(\geq)}{\leq} \left\{ \frac{(r_1-1)!}{k_1^{r_1}} E|X(r_1,n_1,-1,k_1)|^{\alpha_1 p_1} \right\}^{1/p_1} \cdot \left\{ \frac{(r_2-1)!}{k_2^{r_2}} E|X(r_2,n_2,-1,k_2)|^{\alpha_2 p_2} \right\}^{1/p_2}$$

where $p_1 = \dfrac{r_1 - r_2}{1 - r_2} = \dfrac{k_1 - k_2}{n - k_2} \gtrless 1$, if both ratios are not of the type $\frac{0}{0}$; otherwise, p_1 has to be chosen in the prescribed way.

ii)
$$\frac{1}{k} E|X(1,n,-1,k)|^{\alpha} \underset{(\geq)}{\leq} \left\{ \left(r_1 \binom{n_1}{r_1}\right)^{-1} E|X(r_1,n_1,0,1)|^{\alpha_1 p_1} \right\}^{1/p_1} \cdot \left\{ \left(r_2 \binom{n_2}{r_2}\right)^{-1} E|X(r_2,n_2,0,1)|^{\alpha_2 p_2} \right\}^{1/p_2}$$

where $p_1 = \dfrac{r_1 - r_2}{1 - r_2} = \dfrac{(n_1 - r_1) - (n_2 - r_2)}{k - (1 + n_2 - r_2)} \gtrless 1$, if both ratios are not of the type $\frac{0}{0}$; otherwise, p_1 has to be chosen in the prescribed way.

REMARK 1.1.6.

i) We obtain equality in 1.1.5. i), iff for some $c > 0$ $(m = -1)$

$$|F^{-1}(t)|^{\alpha_1 p_1 - \alpha_2 p_2} = c\,(1-t)^{k_2-k_1}\,(\log \tfrac{1}{1-t})^{r_2-r_1} \quad \text{a.e. on } (0,1)\,.$$

If we assume $F^{-1}(t) > 0$ for all $t \in (0,1)$ (for simplicity), we may consider two cases :

1. $k_1 = k_2$ and $r_1 \neq r_2$. Hence (see the assumptions), $k_1 = k_2 = n$ and we obtain n–th records on the r.h.s.

If $r_1, r_2 \in \mathbb{N}^{\geq 2}$, $r_1 \neq r_2$, then $\frac{r_2 - r_1}{r_2 - 1} \begin{cases} \in (0,1)\,, & r_2 > r_1 \\ < 0\,, & r_2 < r_1 \end{cases}$ and hence $p_1 < 1$;

i.e. the inverse Hölder inequality is used.

We find equality iff for some $c > 0$

$$F^{-1}(t) = \left(c\,(\log \tfrac{1}{1-t})^{r_2-r_1}\right)^{1/\tilde{\alpha}} \quad \text{a.e. on } (0,1) \text{ with } \frac{\tilde{\alpha}}{r_2-r_1} > 0\,.$$ Thus F is a

Weibull distribution function : $F(x) = 1 - \exp(-(\frac{x}{c})^{\tilde{\alpha}})^{1/(r_2-r_1)})$.

2. $r_1 = r_2$ and $k_1 \neq k_2$. Hence (see the assumptions) $r_1 = r_2 = 1$ and, on the r.h.s., we obtain moments of a minimum of k_1 or k_2 random variables distributed according to F (cf. marginal density with $r = 1$, $m = -1$).
We find equality iff for some $c > 0$

$$F^{-1}(t) = \left(c\,(1-t)^{k_2-k_1}\right)^{1/\tilde{\alpha}} \quad \text{a.e. on } (0,1) \text{ with } \frac{\tilde{\alpha}}{k_2-k_1} > 0\,.$$ Thus F is a

Pareto distribution function : $F(x) = 1 - (\frac{x}{c})^{\tilde{\alpha}})^{1/(k_2-k_1)}$, $x \geq c^{1/\tilde{\alpha}}$.

ii) We obtain equality in 1.1.5. ii), iff for some $c > 0$ $(m = 0)$

$$|F^{-1}(t)|^{\alpha_1 p_1 - \alpha_2 p_2} = c\,(1-t)^{(n_2-r_2)-(n_1-r_1)}\,t^{r_2-r_1} \quad \text{a.e. on } (0,1)\,.$$

Assuming $F^{-1}(t) > 0$ for all $t \in (0,1)$ we consider two cases :

1. $n_2 - r_2 = n_1 = r_1$ and $r_1 \neq r_2$. Hence (see the assumptions),

$k = n_2 - r_2 + 1$ and we have some first k–th record on the l.h.s.

We observe equality iff for some $c > 0$

$$F^{-1}(t) = \left(c\, t^{r_2-r_1}\right)^{1/\tilde{\alpha}} \quad \text{a.e. on } (0,1) \text{ with } \frac{\tilde{\alpha}}{r_2-r_1} > 0 .$$ Thus F is power function distribution function : $F(x) = (\frac{x}{c}^{\tilde{\alpha}})^{1/(r_2-r_1)}$, $x \in (0, c^{1/\tilde{\alpha}})$.

2. $r_1 = r_2$ and $n_1 \neq n_2$. Hence $r_1 = r_2 = 1$ and, on the r.h.s., we find moments of minima of n_1 and n_2 random variables, respectively, distributed according to F.
We observe equality iff for some $c > 0$

$$F^{-1}(t) = \left(c\,(1-t)^{n_2-n_1}\right)^{1/\tilde{\alpha}} \quad \text{a.e. on } (0,1) \text{ with } \frac{\tilde{\alpha}}{n_2-n_1} < 0 .$$ Thus F is a Pareto distribution function : $F(x) = 1 - (\frac{x}{c}^{\tilde{\alpha}})^{1/(n_2-n_1)}$, $x \geq c^{1/\tilde{\alpha}}$.

Choosing the parameters appropriately, characterization results, e.g. for exponential and uniform distributions, arise by means of moments of o OS's and record values.

COROLLARY 1.1.7. With the notations and under the assumptions of Theorem 1.1.1., assuming $F^{-1}(t) > 0$ for all $t \in (0,1)$ and $r_1, r_2 > 1$, $r_1 \neq r_2$, we find :

i)
$$E|X(1,n,0,1)|^{\alpha} \geq \left\{(r_1-1)!\ E|X(r_1,n_1,-1,n)|^{-(r_1-1-\alpha)}\right\}^{(r_2-1)/(r_2-r_1)} \cdot \left\{(r_2-1)!\ E|X(r_2,n_2,-1,n)|^{-(r_2-1-\alpha)}\right\}^{-(r_1-1)/(r_2-r_1)}$$

with equality iff for some $\lambda > 0$: $F(x) = 1 - \exp(-x/\lambda)$, $x > 0$.

ii)
$$\tfrac{1}{k} E|X(1,n,-1,k)|^{\alpha} \geq \left\{\left(r_1\binom{k+r_1-1}{r_1}\right)^{-1} E|X(r_1,k+r_1-1,0,1)|^{-(r_1-1-\alpha)}\right\}^{(r_2-1)/(r_2-r_1)} \cdot \left\{\left(r_2\binom{k+r_2-1}{r_2}\right)^{-1} E|X(r_2,k+r_2-1,0,1)|^{-(r_2-1-\alpha)}\right\}^{-(r_1-1)/(r_2-r_1)}$$

with equality iff for some $a > 0$: $F(x) = x/a$, $x \in (0,a)$.

1.2. An Inequality between Two Moments

A particular case has not been considered yet. In Theorem 1.1.1. one can choose $\alpha_1 = 0$ and $\alpha_2 = \alpha$ to obtain inequalities between two moments of g OS's. In this case we may weaken the assumptions. E.g., the condition imposed on the exponents p_1 and p_2 in Hölder's inequality may be dropped, if two additional assumptions are made depending on m (see the remarks to Theorem 1.2.1.). Corresponding results in the case of o OS's and records, examples and characterizations are presented in Kamps (1991a). Moreover, other inequalities are derived based on integral inequalities of Jensen (e.g. Mitrinović 1970), Diaz, Metcalf (see Mitrinović 1970, p 63) and Pólya, Szegö (1964, p 57).

Theorem 1.2.1. (Kamps 1991a for $m \in \{0,-1\}$)

Let F be the underlying, non-degenerate distribution function of the appearing g OS's.

Let $\alpha \in \mathbb{R}$, $p \in \mathbb{R}\backslash\{0,1\}$, $r,r',n,n' \in \mathbb{N}$, $k,k' \geq 1$, $r \leq n$, $r' \leq n'$,

$\tilde{m} = (m_1,\dots,m_{n-1})' \in \mathbb{R}^{n-1}$, $\tilde{m}' = (m_1',\dots,m_{n'-1}')' \in \mathbb{R}^{n'-1}$,

satisfying the regularity conditions $m_1 = \dots = m_{r-1} = m_1' = \dots = m_{r'-1}' = m$,

$$b_i = k + n - i + M_i \geq 1, \; i = 1,\dots,n-1,$$

$$b_i' = k' + n' - i' + M_{i'}' \geq 1, \; i' = 1,\dots,n'-1 \quad (M_{i'}' = \sum_{j=i'}^{n'-1} m_j')$$

and $F^{-1}(t) \neq 0$ for all $t \in (0,1)$, if $\alpha < 0 \vee \alpha p < 0$.

Moreover, let $R = \frac{rp - r'}{p-1} > 0$ and

$$C_m = \begin{cases} \dfrac{b_r p - b_{r'}'}{p-1} > 0 & \text{, if } m \geq -1 \\ \dfrac{b_r p - b_{r'}'}{p-1} + (m+1)\left(\dfrac{rp - r'}{p-1} - 1\right) > 0, & \text{if } m < -1 \end{cases}.$$

Then we find with $p > 1$ ($0 \neq p < 1$):

$$E|X(r,n,\tilde{m},k)|^{\alpha} \underset{(\geq)}{\leq} \frac{c_{r-1}(\tilde{m},k)}{(r-1)!} A_m^{1-1/p} \left(\frac{c_{r'-1}(\tilde{m}',k')}{(r'-1)!}\right)^{-1/p} \left(E|X(r',n',\tilde{m}',k')|^{\alpha p}\right)^{1/p}$$

where
$$A_m = \begin{cases} \left(\frac{1}{m+1}\right)^R \; B\left(\frac{C_m}{m+1}, R\right), & m > -1 \\ C_{-1}^{-R} \; \Gamma(R), & m = -1 \\ \left(-\frac{1}{m+1}\right)^R B\left(-\frac{C_m}{m+1}, R\right), & m < -1 \end{cases} .$$

Equality holds iff for some constant $c > 0$

$$|F^{-1}(t)|^{\alpha p} = c\,(1-t)^{\frac{p}{p-1}((k+n-r+M_r)-(k'+n'-r'+M'_{r'}))} (g_m(t))^{\frac{p}{p-1}(r-r')} \quad \text{a.e. on } (0,1).$$

Proof By analogy with 1.1.1., the proof is directly obtained using the Hölder inequality ($p > 1$) or the inverse Hölder inequality ($0 \neq p < 1$).

The above theorem contains a variety of interesting special cases. Examples with respect to o OS′s and records are shown in Kamps (1991a). Choosing the parameters appropriately, characterization theorems for those distributions result which are mentioned in Theorem 1.1.1. The conditions of positivity of R and C_m may be interpreted as conditions imposed on p or as restrictions with respect to the other parameters.

In the case of o OS′s we have

$$C_0 = \frac{(n-r)p - (n'-r')}{p-1} + 1 ;$$

i.e., if $p > 1$ then, e.g., the assumption $n - r \geq n' - r'$, $r \geq r'$ is sufficient. In Kamps (1991a), $n - r = n' - r'$ with $n' \leq n$ is chosen leading to a characterization of power function distributions.

In the case of 1–records we have

$$C_{-1} = \frac{kp - k'}{p-1} = 1 ;$$

i.e., the assumption $r \geq r'$ is sufficient for $R > 0$, if $p > 1$, and it leads to characterization results for Weibull distributions.

The results of Lin (1988a) (cited above) are special cases of Theorem 1.2.1. putting $\alpha = 1$, $p = 2$, $r' = r - 1$ and

$m_1 = \dots = m_{n-1} = m_1' = \dots = m_{n'-1}' = 0$, $n' = n - 1$, $k = k' = 1$ for o OS′s

and

$m_1 = \dots = m_{n-1} = m_1' = \dots = m_{n'-1}' = -1$, $k = k' = 1$ for 1–records.

Moreover, we obtain general bounds for expectations of maxima and n–th record values.

We find ($\alpha = 1$, $p = 2$) :

$$E|X_{n,n}| \le n/(2n-1)^{1/2}\,(EX^2)^{1/2}$$

and

$$E|X_{L(n)}| \le ((2n-2)!)^{1/2}/((n-1)!)\,(EX^2)^{1/2}\,.$$

In contrast to the examples in the introductory part of this chapter we do not assume centred random variables (cf. remarks in Gajek, Gather 1989).

Taking up Lin's (1988a) remark again, we may interpret Theorem 1.2.1. in this way : Any distribution which is determined by

$$|F^{-1}(t)|^{\alpha p} = c\,(1-t)^{\frac{p}{p-1}((k+n-r+M_r)-(k'+n'-r'+M'_{r'}))}\,(g_m(t))^{\frac{p}{p-1}(r-r')}$$

can be characterized by means of two special moments (cf. Section III.3.).

Remark

If the existence of some moment of a certain g OS is ensured, Theorem 1.2.1. may be used to show the existence of moments of other g OS's. Grudzień, Szynal (1983, Lemma 1) show, that the existence of the p–th absolute moment ($p > 1$) of a minimum of k random variables with a continuous distribution function F implies the existence of all expectations of k–th records :

If $E|X_{1,k}|^p$ exists for some $p > 1$ and $k \in \mathbb{N}$, then $EX_{L^{(k)}(r)}$ exists for all $r \in \mathbb{N}$.

This result is immediately obtained from Theorem 1.2.1. putting $\alpha = 1$, $r' = 1$.

In the case $k = 1$, i.e. ordinary records, we have Lemma 2.1. of Nagaraja (1978). In this article it is also shown that the existence of $EX_{L(r)}$ and of $E(X\,(\log X)^r)$ are equivalent conditions, if we consider non–bounded and positive random variables. Grudzień, Szynal (1983) state a generalization to k–th records with the existence of $E(X\,(\log X)^r)$ replaced by the existence of $E(X_{1,k}\,(\log X_{1,k})^r)$.

2. The Application of Other Integral Inequalities

The application of Jensen's inequality to moments of g OS's leads to characterizations of power function distributions. In the case of o OS's and records the results are stated in Arnold, Balakrishnan (1989) and in Kamps (1991a).

THEOREM 2.1. Let $\alpha > 0$, $F^{-1}(0) \geq 0$, and $(F^{-1}(t))^{\alpha}$ be convex (concave) , $t \in (0,1)$. Then, concerning the moment of $X(r,n,\tilde{m},k)$ of order α and $m_1 = ... = m_{r-1}$ we have

$$EX^{\alpha}(r,n,\tilde{m},k) \underset{(\leq)}{\geq} (F^{-1}(ET))^{\alpha}$$

where the random variable T is distributed according to the density $\varphi_{r,n}$ (see I.3.2.1.).
We obtain equality iff F is degenerate or given by

$$F(x) = \tfrac{1}{a}(x^{\alpha} - b) , a > 0 , b \geq 0 , x \in (b^{1/\alpha},(a+b)^{1/\alpha}) .$$

PROOF Let the random variable T be distributed according to $\varphi_{r,n}$. Then we have

$$EX^{\alpha}(r,n,\tilde{m},k) = \int_0^1 (F^{-1}(t))^{\alpha} \varphi_{r,n}(t)\, dt = E\,(F^{-1}(T))^{\alpha} \underset{(\leq)}{\geq} (F^{-1}(ET))^{\alpha}$$

with equality iff $(F^{-1}(t))^{\alpha}$ is linear on $(0,1)$: $(F^{-1}(t))^{\alpha} = at + b$.

REMARK An explicit expression of ET is shown in I.3.2.11.

In the case of o OS's and $\alpha = 1$, this result is already stated in Blom (1958, p 68); Arnold, Balakrishnan (1989, p 58) give this reference. OS's are called 'transformed beta–variables' in Blom's monograph and other results for moments of OS's from specific distributions can be found.
Generalizations of this result are stated in Arnold, Balakrishnan (1989, p 59) and can be transferred to g OS's analogously.

If f and g are real valued and quadratic integrable functions on an interval [a,b] and if the condition

$$K_1 \le \frac{g(x)}{f(x)} \le K_2 \quad \text{a.e. with respect to } x \in [a,b],$$

$f(x) \neq 0$ and constants K_1, K_2 is fulfilled, then we have an inequality of Diaz and Metcalf (see Mitrinović 1970, p 63):

$$(K_1 + K_2) \int_a^b f(x)\, g(x)\, dx \ \ge \int_a^b g^2(x)\, dx \ + \ K_1 K_2 \int_a^b f^2(x)\, dx\,.$$

Moreover, equality in this inequality can be characterized. Applying this inequality we obtain

THEOREM 2.2. Let F be the underlying distribution function of the appearing g OS's, F^{-1} continuous on (0,1) and let some $a \in [0,1)$ exist with

$$\begin{cases} F^{-1}(t) = 0 \text{ for all } t \le a \\ F^{-1}(t) > 0 \text{ for all } t > a \end{cases}.$$

Let the regularity conditions of Theorem 1.1.1. be fulfilled with respect to $r, r_i, n, n_i, k, k_i, \tilde{m}, \tilde{m}_i$ and let

$$r = \tfrac{1}{2}(r_1 + r_2)\,,\ k + n + M_r = \tfrac{1}{2}(k_1 + n_1 + M^{(1)}_{r_1} + k_2 + n_2 + M^{(2)}_{r_2})\,.$$

Moreover, let $\alpha, \alpha_1, \alpha_2 > 0$ with $\alpha_1 + \alpha_2 = \alpha$.

If there exist some constants K_1, K_2 such that

$$K_1 \le H(t) = (F^{-1}(t))^{\alpha_1-\alpha_2} (1-t)^{\frac{1}{2}(k_1-k_2+n_1-n_2-(r_1-r_2)+M^{(1)}_{r_1}-M^{(2)}_{r_2})}\, g_m^{\frac{1}{2}(r_1-r_2)}(t) \le K_2$$

is valid for all $t \in (a,1)$, then we have

$$(K_1 + K_2)\, EX^{\alpha}(r,n,\tilde{m},k) \ge \frac{c_r - 1}{c_{r_1} - 1} \frac{(r_1 - 1)!}{(r-1)!}\, EX^{2\alpha_1}(r_1,n_1,\tilde{m}_1,k_1)$$

$$+ K_1 K_2 \frac{c_r - 1}{c_{r_2} - 1} \frac{(r_2 - 1)!}{(r-1)!}\, EX^{2\alpha_2}(r_2,n_2,\tilde{m}_2,k_2)\,.$$

Equality holds iff $H(t) \in \{K_1, K_2\}$ for all $t \in (a,1)$.

The discussion of all those cases in which H(t) is constant for all $t \in (a,1)$ leads to characterizations of special distributions. In its general form, the theorem includes inequalities, e.g., between o OS's and records (cf. remark before 1.1.5.).

In the special case

$$m_1 = \ldots = m_{n-1} = m_1^{(1)} = \ldots = m_{n_1-1}^{(1)} = m_1^{(2)} = \ldots = m_{n_2-1}^{(2)} = m \text{, say,}$$

H(t) simplifies to

$$H(t) = (F^{-1}(t))^{\alpha_1-\alpha_2}(1-t)^{\frac{1}{2}(k_1-k_2+(n_1-n_2-(r_1-r_2))(m+1))}\, g_m^{\frac{1}{2}(r_1-r_2)}(t)\,.$$

The case $m = 0$ (o OS's) is discussed in Kamps (1991a) and, e.g., power function, Pareto and Burr XII distributions can be characterized. The following corollary shows the results in subclasses of g OS's with respect to an arbitrary m.

COROLLARY 2.3. Let the assumptions of Theorem 2.2. be given and

$$m_1 = \ldots = m_{n-1} = m_1^{(1)} = \ldots = m_{n_1-1}^{(1)} = m_1^{(2)} = \ldots = m_{n_2-1}^{(2)} = m\,.$$

If $\alpha_1 > \alpha_2$, we obtain equality in Theorem 2.2., iff one of the following conditions is satisfied for $c \in \{K_1,K_2\}$, $c > 0$:

i) $r_1 = r_2$, $k_1 - k_2 + (n_1 - n_2)(m + 1) = 0$

and F is degenerate at $c^{1/(\alpha_1-\alpha_2)}$;

ii) $r_1 = r_2$, $k_1 - k_2 + (n_1 - n_2)(m + 1) > 0$

and F is a Pareto distribution function,

i.e. $F(x) = 1 - \left(c/(x^{\alpha_1-\alpha_2})\right)^{2/(k_1-k_2+(n_1-n_2)(m+1))}$, $x \in (c^{1/(\alpha_1-\alpha_2)}, \infty)$;

iii) $r_1 < r_2$, $k_1 - k_2 + (n_1 - r_1 - (n_2 - r_2))(m + 1) = 0$

and F is given by

$$F(x) = \begin{cases} 1 - \left(1 - (m+1)(c/(x^{\alpha_1-\alpha_2}))^{2/(r_1-r_2)}\right)^{1/(m+1)} & , m \neq -1 \\ 1 - \exp\left(-\,(c/(x^{\alpha_1-\alpha_2}))^{2/(r_1-r_2)}\right) & , m = -1 \end{cases}$$ with

$$\begin{cases} x \in (0\,,\,(c(m+1))^{(r_1-r_2)/2})^{1/(\alpha_1-\alpha_2)}) & , m > -1 \\ x \in (0\,,\,\infty) & , m \le -1 \end{cases}$$

($m = 0$: power function distributions, $m < -1$: Burr XII distributions, $m = -1$: Weibull distributions);

iv) $r_1 < r_2\,,\; k_1 - k_2 + (n_1 - r_1 - (n_2 - r_2))(m + 1) > 0$

and F is given by

$$F^{-1}(t) = \left(c\,(1-t)^{-\frac{1}{2}(k_1-k_2+(n_1-n_2-(r_1-r_2))(m+1))}\, g_m^{-\frac{1}{2}(r_1-r_2)}(t)\right)^{1/(\alpha_1-\alpha_2)} .$$

Some other special cases of Theorem 2.2. are shown in the following corollary.

COROLLARY 2.4. Under the assumptions of Corollary 2.3. we have :

i) $(K_1 + K_2)\, EX^{\alpha}(r,n,m,k) \ge EX^{2\alpha_1}(r,n,m,k) + K_1\, K_2\, EX^{2\alpha_2}(r,n,m,k)$

with equality, if F is degenerate; $K_1^{1/(\alpha_1-\alpha_2)}$ and $K_2^{1/(\alpha_1-\alpha_2)}$ are bounds for the support of F.

ii) $(K_1 + K_2)\, EX^{\alpha}(r,n,m,k) \ge \frac{k\,+\,(n-r)\,(m+1)}{k\,+\,n(m+1\,)}\, EX^{2\alpha_1}(r,n+1,m,k)$

$$+ K_1\, K_2\, \frac{k\,+\,(n-1)(m+1\,)}{k\,+\,(n-r-1)\,(\,m+1)}\, EX^{2\alpha_2}(r,n-1,m,k)$$

with equality, if $m > -1$, and F is a Pareto distribution function.

iii) $(K_1 + K_2)\, EX^{\alpha}(r,n,m,k) \ge \frac{k\,+\,(n-1)(m+1)}{r\,-\,1}\, EX^{2\alpha_1}(r-1,n-1,m,k)$

$$+ K_1\, K_2\, \frac{r}{k\,+\,n(m+1)}\, EX^{2\alpha_2}(r+1,n+1,m,k)$$

with equality, e.g., for power function and Weibull distributions (see 2.3. iii)).

iv) $(K_1 + K_2)\, EX^{\alpha}(r,n,m,k) \ge \frac{k\,+\,(n-r)(m+1)}{r\,-\,1}\, EX^{2\alpha_1}(r-1,n,m,k)$

$$+ K_1\, K_2\, \frac{r}{k+(n-r-1)(m+1)}\, EX^{2\alpha_2}(r+1,n,m,k)$$

with equality, if $m > -1$, and F is a Burr XII distribution function of the form $F(x) = 1 - \left(1 + \frac{m+1}{c} x^{\alpha_1 - \alpha_2}\right)^{-1/(m+1)}$, $x \in (0,\infty)$, $c \in \{K_1, K_2\}$.

Pólya, Szegö (1964, p 57) present the following integral inequality :

Let $f(x)$ be defined for $0 < x < 1$, nondecreasing, nonnegative, but not identically equal to 0. Moreover, let $0 < a < b$.
If all appearing integrals exist, we have

$$1 - \left(\frac{a - b}{a + b + 1}\right)^2 \le \frac{\left(\int_0^1 x^{a+b} f(x)\, dx\right)^2}{\int_0^1 x^{2a} f(x)\, dx \int_0^1 x^{2b} f(x)\, dx} < 1 .$$

The second inequality is well–known. In the first inequality we have equality iff f is constant.

The first inequality thus leads to inequalities for moments of g OS's based on a distribution with support (0,1), if the restrictive conditions imposed on the density $f^{X(r,n,\tilde{m},k)}$ are fulfilled.

THEOREM 2.5. Let $X(r,n,\tilde{m},k)$ be some g OS with an underlying distribution function F and support (0,1) and monotonically increasing density function $f^{X(r,n,\tilde{m},k)}$.
If the appearing moments exist and if $0 < a < b$, we have :

$$\left(EX^{a+b}(r,n,\tilde{m},k)\right)^2 \ge \frac{(2a + 1)(2b + 1)}{(a + b + 1)^2} EX^{2a}(r,n,\tilde{m},k) \cdot EX^{2b}(r,n,\tilde{m},k) .$$

Equality holds iff $f^{X(r,n,\tilde{m},k)}$ is constant on (0,1) ; i.e. $f^{X(r,n,\tilde{m},k)}$ is the density function of the uniform distribution on (0,1) .

REMARK The conditions of Theorem 2.5. can be fulfilled since, e.g., the density function of the maximum of n random variables from the power function distribution with $F(x) = x^a$ ($f^{X(n,n,0,1)}(x) = a\, n\, x^{na-1}$) increases strictly, if $n\, a > 1$.

A constant density function $f^{X(r,n,\tilde{m},k)}$ appears, e.g., if $r = n = 2$, $m = 0$, $k = 1$, $F(x) = x^{1/2}$.

I.e., the maximum of two random variables distributed according to $F(x) = x^{1/2}$ possesses a uniform distribution on $(0,1)$. More generally, we have :

Just those distributions lead to equality for which in the case $m = 0$ the distribution function of the r–th o OS equals a uniform distribution function on $(0,1)$.

The distribution function of $X(r,n,0,1)$ may be expressed via the incomplete beta–integral (see David 1981, p 8) :

$$F^{X(r,n,0,1)}(x) = \frac{1}{B(r,n-r+1)} \int_0^{F(x)} t^{r-1} (1-t)^{n-r}\, dt .$$

If F is supposed to be continuous, then ($F^{X(r,n,0,1)}(x) = x$)

$$F^{-1}(x) = \frac{1}{B(r,n-r+1)} \int_0^{x} t^{r-1} (1-t)^{n-r}\, dt .$$

These distributions on the interval $(0,1)$ form a subclass of the family of distributions defined in Kamps (1991b) (see III.2.1.1.).

E.g., the following distributions are contained :

$r = 1$: $\quad F(x) = 1-(1-x)^{1/n}$, $x \in (0,1)$, and

$r = n$: $\quad F(x) = x^{1/n}$, $x \in (0,1)$.

3. A Special Application of Cauchy's Inequality

In Theorem 1.1., we cite results of Lin (1988a) which are then generalized via Hölder's inequality and transferred to g OS's. Two other theorems in this article are motivated by the fact that we do not characterize arbitrary uniform or exponential distributions. However, if an additional moment is considered, we obtain the desired results. Theorem 3.1. shows the generalization to g OS's and, by this, how to generalize Lin's inequalities which are deduced by means of Cauchy's inequality.

THEOREM 3.1. Let $X(r,n,\tilde{m},k)$, $X(1,1,\tilde{m}_1,k)$ and $X(1,1,\tilde{m}_2,2k-1)$ be g OS's based on F with existing moments of orders α and 2α, respectively.

Moreover, let $a \in \mathbb{R}$, $n \geq 2$, $m_1 = ... = m_{r-1} = m$ and $\begin{cases} n-r+M_r \geq 0 & , m \geq -1 \\ n+M_1 \geq 1 & , m < -1 \end{cases}$.

Then we find :

$$\left(EX^{\alpha}(r,n,\tilde{m},k) - a\, EX^{\alpha}(1,1,\tilde{m}_1,k)\right)^2$$

$$\leq \frac{1}{2k-1} \int_0^1 \left(\frac{c_{r-1}}{(r-1)!} (1-t)^{n-r+M_r} g_m^{\,r-1}(t) - a\,k\right)^2 dt \cdot EX^{2\alpha}(1,1,\tilde{m}_2,2k-1) .$$

Equality holds iff

$$(F^{-1}(t))^{\alpha} (1-t)^{k-1} = d \left(\frac{c_{r-1}}{(r-1)!} (1-t)^{n-r+M_r} g_m^{\,r-1}(t) - a\,k\right)$$

for some constant d .

PROOF The application of Cauchy's inequality (e.g. Mitrinović 1970, p 43) leads to

$$\left(EX^{\alpha}(r,n,\tilde{m},k) - a\, EX^{\alpha}(1,1,\tilde{m}_1,k)\right)^2$$

$$= \left(\int_0^1 (F^{-1}(t))^{\alpha} (1-t)^{k-1} \left(\frac{c_{r-1}}{(r-1)!} (1-t)^{n-r+M_r} g_m^{\,r-1}(t) - a\,k\right) dt \right)^2$$

$$\leq \int_0^1 (F^{-1}(t))^{\alpha} (1-t)^{2(k-1)}\, dt \int_0^1 \left(\frac{c_{r-1}}{(r-1)!} (1-t)^{n-r+M_r} g_m^{\,r-1}(t) - a\,k\right)^2 dt .$$

REMARK Since first g OS's appear in the above theorem, the vectors $\tilde{m}_1$ und $\tilde{m}_2$ may be different from $\tilde{m}$.

The integral appearing on the r.h.s. can be expressed explicitly in our situation (see e.g. Gröbner, Hofreiter 2, 1961, p 18,74). Here, we omit this representation as well as representations of the characterized distribution functions.

The special case $r = n$ is shown in the following corollary. Then we note the corresponding distribution functions for $r = n$ and $k = 1$.

COROLLARY 3.2. Under the conditions of Theorem 3.1. and with $r = n$ (hence $m_1 = ... = m_{n-1} = m \geq -1$) we observe in the case

$m > -1$: $$\left(EX^{\alpha}(n,n,m,k) - a\,EX^{\alpha}(1,1,\tilde{m}_1,k)\right)^2$$
$$\le \frac{1}{2k-1}\left(\left(\frac{c_{n-1}}{(n-1)!}\right)^2 \left(\frac{1}{m+1}\right)^{2n-1} (2n-2)!\,\frac{\Gamma(\frac{1}{m+1})}{\Gamma(\frac{1}{m+1}+2n-1)}\right.$$
$$\left. -2\,a\,k\,c_{n-1}\left(\frac{1}{m+1}\right)^n \frac{\Gamma(\frac{1}{m+1})}{\Gamma(\frac{1}{m+1}+n)} + a^2k^2\right) EX^{2\alpha}(1,1,\tilde{m}_2,2k-1)\,,$$

$m = -1$: $$\left(EX^{\alpha}(n,n,-1,k) - a\,EX^{\alpha}(1,1,\tilde{m}_1,k)\right)^2$$
$$\le \left(\frac{k^{2n}}{((n-1)!)^2}(2n-2)! - 2\,a\,k^{n+1} + a^2k^2\right)\cdot EX^{2\alpha}(1,1,\tilde{m}_2,2k-1)\,.$$

Remark Lin (1988a) considers $m = 0$ and $m = -1$ if $k = 1$. Then the constants appearing on the r.h.s. are given by

$$(a-1)^2 + \frac{(n-1)^2}{2n-1} \quad\text{and}\quad (a-1)^2 + \frac{(2n-2)!}{((n-1)!)^2} - 1\,,$$

respectively.

If $k = 1$ is assumed in Corollary 3.2., then the distributions which are characterized by equality are determined by

$$(F^{-1}(t))^{\alpha} = d\left(\frac{c_{n-1}}{(n-1)!}\,g_m^{n-1}(t) - a\right).$$

I.e., $$F(x) = g_m^{-1}\left(\left[\frac{(n-1)!}{c_{n-1}}\left(\frac{1}{d}x^{\alpha} + a\right)\right]^{1/(n-1)}\right)$$

with $$g_m^{-1}(x) = \begin{cases} 1-(1-(m+1)\,x)^{1/(m+1)}\,, & m > -1 \\ 1 - e^{-x}\,, & m = -1 \end{cases}$$

In particular, the results of Lin (1988a) for $m = 0, -1$ are included. We refer to Lin's remarks concerning the supports of the probability distributions.

Chapter V

Reliability Properties of Generalized Order Statistics

In reliability theory, classes of distributions are considered which describe the life–length of components or systems. I.e., the probability of failure up to a time $t \geq 0$ is modelled. There are numerous articles concerning the analysis of such families of distributions (see Barlow, Proschan 1975, 1981, Patel 1983, Basu 1988). We restrict ourselves to some articles dealing with o OS′s and record values.
If a distribution with distribution function F possesses a density function f, then the failure rate (hazard rate) λ is defined by

$$\lambda(t) = \frac{f(t)}{1 - F(t)}, \; t \geq 0,$$

leading to the examination of classes of distributions having increasing failure rate (IFR property) or decreasing failure rate (DFR property).

Interpreted as life–length distributions, we assume $F^{-1}(0) \geq 0$ throughout this chapter.

The exponential distribution stands out in this context since it has a constant failure rate. The exponential distribution can be characterized by this property (e.g. Azlarov, Volodin 1986, p 15).

Moreover, classes of distributions are formed via monotonicity of $-\frac{1}{t}\log(1 - F(t))$ with $t \geq 0$ (IFRA: increasing failure rate average; DFRA: decreasing failure rate average).

When considering such classes of distributions, we are interested in knowing whether we have certain transmission properties at hand. Does the IFR property of some distribution imply the IFR property of a given statistic? In this context o OS′s are considered (see Nagaraja 1990), since the r–th OS $X_{r,n}$ from a sample of size n represents the life–length of a $(n-r+1)$–out–of–n–system consisting of n components of the same kind with independent life–lengths ($r = 1$: series system; $r = n$: parallel system) (cf. Section I.1., Barlow, Proschan 1975, p 59).
In the case of o OS′s, the IFR (IFRA) property of some distribution function F implies the corresponding property of an arbitrary OS based on F (e.g. Barlow, Proschan 1965, p 36 and 1975, p 108).

Other classes of distributions are, e.g., defined via the validity of the following inequality :

$$1 - F(x + y) \underset{(\geq)}{\leq} (1 - F(x))(1 - F(y)) \quad \text{for all} \quad x,y \geq 0$$

(NBU: new better than used; NWU: new worse than used)
or, if the expectation

$$\mu = \int_0^\infty (1 - F(t))\, dt$$

exists, via the monotonicity of

$$\frac{1}{1 - F(x)} \int_x^\infty (1 - F(t))\, dt$$

(IMRL: increasing mean residual life; DMRL: decreasing mean residual life).

Concerning interpretations and examples, we refer to the literature on reliability theory (see Barlow, Proschan 1975, 1981, Basu 1988). E.g., it is shown that the following implications hold :

$$\text{'IFR} \Longrightarrow \text{IFRA} \Longrightarrow \text{NBU'} \quad \text{and} \quad \text{'DFR} \Longrightarrow \text{DFRA} \Longrightarrow \text{NWU'} .$$

In the two sections of this chapter, we deal with the transmission of aging properties such as IFR or DFR property and we consider partial ordering of g OS's.

1. The Transmission of Aging Properties

The transmission of the IFR property from some underlying distribution to the OS's (Barlow, Proschan 1975, p 108) is taken up by Takahasi (1988) and generalized to the transmission of the IFR property from the r–th OS to the (r+1)–th OS (see Remark 1.6.).
The DFR property of some distribution does not necessarily imply the DFR property of the corresponding OS's (e.g. Barlow, Proschan 1965, Patel 1983). Takahasi (1988) shows that the DFR property of the r–th OS implies the DFR property of the (r–1)–th OS; but on the other hand, there is no distribution such that all corresponding OS's possess the DFR property.
Nagaraja (1990) generalizes the known results with respect to other criteria used in reliability theory and to other neighbouring OS's. If the r–th OS in a sample of size n

possesses the IFR (IFRA, NBU, DMRL) property, so do $X_{r+1,n}$, $X_{r,n-1}$, $X_{r+1,n+1}$. Under the restriction $r \leq \frac{n+3}{2}$, the assertions remain valid for $X_{r+1,n+2}$ with respect to the first three criteria. Analogously, the DFR (DFRA, NWU) property of $X_{r,n}$ implies the corresponding property of the OS's $X_{r-1,n}$, $X_{r,n+1}$, $X_{r-1,n-1}$. Restricting to $r \leq \frac{n+1}{2}$, the assertions remain valid for $X_{r-1,n-2}$.

The above results are described and discussed in detail by Nagaraja (1990).

The results may be illustrated in the triangular scheme of the OS's (If X_{r_0,n_0} possesses the IFR (DFR) property, then the same property holds for all OS's belonging to the hatched area.) :

IFR Property

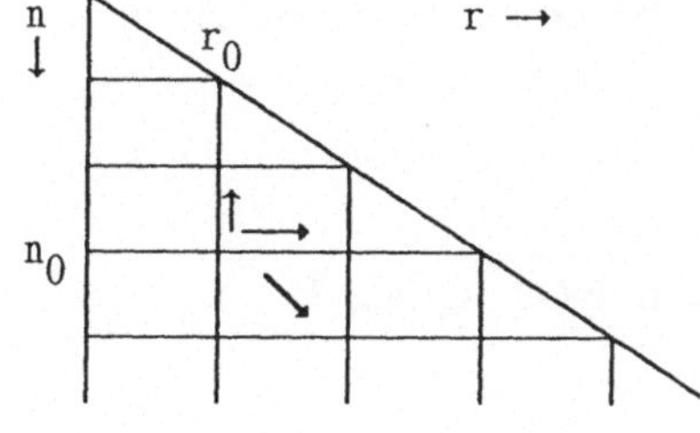

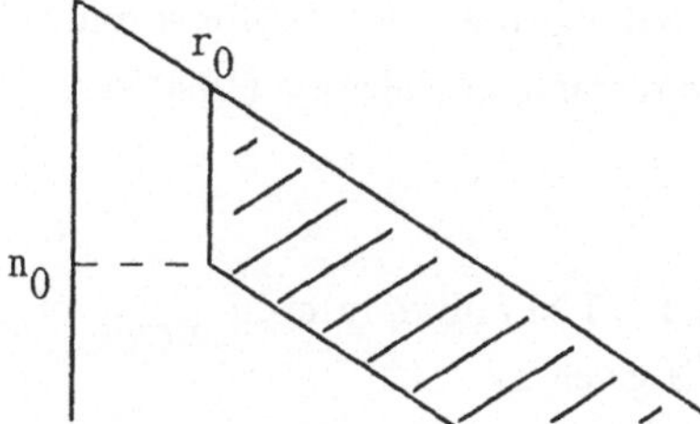

DFR Property

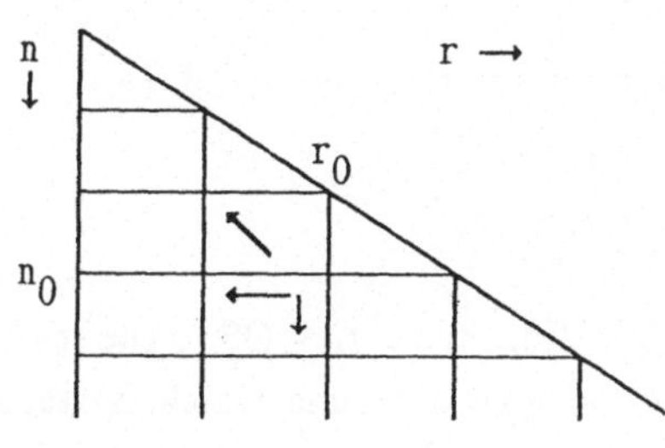

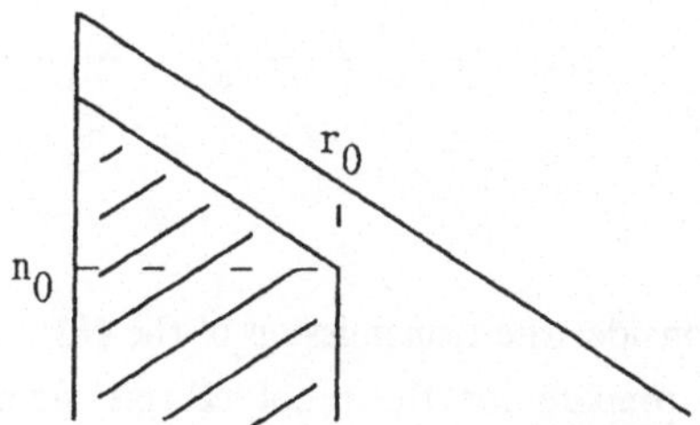

Record values are also used in models of reliability theory. Kochar (1990) points out that records are closely connected with occurrence times of nonhomogeneous Poisson processes and refers to surveys of Ascher, Feingold (1984) and Gupta, Kirmani (1988) (cf. Section I.1.4.). By analogy with the transmission of the IFR property in the case of o OS's it is shown in the latter article that the IFR property of the r–th record is ensured by the IFR property of the underlying distribution. Kochar (1990) generalizes this result in the sense of Takahasi (1988) using the same argument. Thus, the IFR property of some record is transmitted to the following one and the DFR property of some record is transmitted to the previous one. Moreover, if we consider a strictly increasing distribution function, it is not possible that all records possess the DFR property. Gupta, Kirmani (1988) and Kochar (1990) also consider ordering of records and record differences and present some results by analogy with results in the case of o OS's (e.g. Barlow, Proschan 1975, p 108, see Section 2.).

The important results of Takahasi (1988), Kochar (1990) and Nagaraja (1990) are proved in this section for g OS's which are based on an absolutely continuous distribution function F; i.e., the cited assertions for o OS's and record values are particular cases. Thus, the above diagrams illustrate the case of g OS's as well.
The results are stated concerning the IFR and DFR property and may then be extended to IFRA and NBU properties or DFRA and NWU properties as pointed out in 1.16. and noticing the remarks of Nagaraja (1990, p 312/3).

Remark 1.1. The failure rate $h_{r,n,\tilde{m},k}$ of the r–th g OS $X(r,n,\tilde{m},k)$, $m_1 = \ldots = m_{r-1} = m$, say, is given by

$$h_{r,n,\tilde{m},k}(x) = \frac{f^{X(r,n,\tilde{m},k)}(x)}{1 - F^{X(r,n,\tilde{m},k)}(x)}$$

$$= \frac{1}{(r-1)!}\,\frac{f(x)}{1-F(x)}\,\frac{g_m^{r-1}(F(x))}{\sum_{j=0}^{r-1} \frac{1}{j!\,c_{r-j-1}}\, g_m^{j}(F(x))}, \quad x < F^{-1}(1).$$

First we consider the transmission of the IFR property from the r–th g OS to the (r+1)–th g OS. To prepare for the proof of this assertion we give a lemma which is stated in Takahasi (1988) in the case of o OS's ($m = 0$, $k = 1$) and in a modified version (see Remark 1.4) and which is stated in Kochar (1990) in the case of records. The results are obtained by means of differentiation. In the case of g OS's it is not suitable to use differentiation. However, using the definition of a monotone function we directly obtain the assertion including the results for o OS's and records as particular cases.

LEMMA 1.2. The function η_1 with

$$\eta_1(t) = \frac{\sum\limits_{j=0}^{r-1} \frac{1}{j!\, c_{r-j-1}}\, t^{j+1}}{\sum\limits_{j=0}^{r} \frac{1}{j!\, c_{r-j}}\, t^{j}}, \quad r \le n-1,$$

increases strictly in $t \ge 0$ and has range $[0,r)$.

PROOF Let t_1, t_2 with $t_2 > t_1 > 0$ be arbitrary. Then we have

$\eta_1(t_1) < \eta_1(t_2)$ iff

$$A = \left(\sum_{j=0}^{r-1} \frac{1}{j!\, c_{r-j-1}}\, t_1^{j+1}\right)\left(\sum_{j=0}^{r} \frac{1}{j!\, c_{r-j}}\, t_2^{j}\right) - \left(\sum_{j=0}^{r-1} \frac{1}{j!\, c_{r-j-1}}\, t_2^{j+1}\right)\left(\sum_{j=0}^{r} \frac{1}{j!\, c_{r-j}}\, t_1^{j}\right) < 0 .$$

Now,

$$A = \left(\sum_{j=1}^{r} \frac{1}{(j-1)!\, c_{r-j}}\, t_1^{j}\right)\left(\sum_{j=0}^{r} \frac{1}{j!\, c_{r-j}}\, t_2^{j}\right) - \left(\sum_{j=1}^{r} \frac{1}{(j-1)!\, c_{r-j}}\, t_2^{j}\right)\left(\sum_{j=0}^{r} \frac{1}{j!\, c_{r-j}}\, t_1^{j}\right)$$

$$= \frac{1}{c_r} \sum_{j=1}^{r} \frac{1}{(j-1)!\, c_{r-j}}\,(t_1^j - t_2^j) + \sum_{i=1}^{r}\sum_{j=1}^{i-1} \frac{i-j}{i!\, c_{r-i}\, j!\, c_{r-j}}\, t_1^j\, t_2^j\,(t_1^{i-j} - t_2^{i-j}) < 0 ;$$

i.e. η_1 increases strictly.

REMARKS 1.3.

i) The lemma remains valid in the case $r = n$. The constant c_n is not defined if $j = 0$; however, it is obvious that then an arbitrary $c_n > 0$ may be chosen.

ii) In particular, the proof shows that dropping the constant term $1/c_r$ (in the denominator) implies strict monotonicity of the function η with

$$\eta(t) = \frac{\sum\limits_{j=0}^{r-1} \frac{1}{j!\, c_{r-j-1}}\, t^{j+1}}{\sum\limits_{j=1}^{r} \frac{1}{j!\, c_{r-j}}\, t^{j}} = \frac{\sum\limits_{j=0}^{r-1} \frac{1}{j!\, c_{r-j-1}}\, t^{j+1}}{\sum\limits_{j=0}^{r-1} \frac{1}{(j+1)!\, c_{r-j-1}}\, t^{j+1}} .$$

REMARK 1.4. The result of Takahasi (1988) is not directly included in Lemma 1.2., since $c_{r-j} = \frac{n!}{(n-r+j-1)!}$ in the case of o OS's.

Applying a certain combinatorial identity (see Gould 1972, Lemma A.2.1.) together with $f(t) = \frac{t}{1+t}$ yields:

$$\sum_{j=0}^{r} \frac{1}{j!\, c_{r-j}} f^j(t) = \frac{(n-r-1)!}{n!} \sum_{j=0}^{r} \binom{n-r+j-1}{j} f^j(t) = \frac{(n-r-1)!}{n!} (1+t)^{-n} \sum_{j=0}^{r} \binom{n}{j} t^j ;$$

i.e. $\eta_1(f(t)) = (n-r) \dfrac{\sum_{j=0}^{r-1} \binom{n}{j} t^{j+1}}{\sum_{j=0}^{r} \binom{n}{j} t^j}$ (cf. Takahasi 1988).

The following theorems which are proved using Lemma 1.2., contain the results of Takahasi (1988) for o OS's and of Kochar (1990) for records. The extension to k–th record values is obvious when considering η as a function of $k \cdot t$.

THEOREM 1.5. Let $r \leq n-1$, $m_1 = \ldots = m_r = m$. Then we find :
If $X(r,n,\tilde{m},k)$ possesses the IFR property, so does $X(r+1,n,\tilde{m},k)$.

PROOF Let $t = t(x) = g_m(F(x))$.
Thus, t is an increasing function with respect to x and $\dfrac{h_{r+1,n,\tilde{m},k}(x)}{h_{r,n,\tilde{m},k}(x)} = \frac{1}{r}\,\eta_1(t)$.

Now, η_1 is increasing with respect to t, so $h_{r+1,n,\tilde{m},k}(x)$ increases since $h_{r,n,\tilde{m},k}(x)$ increases.

REMARK 1.6. Let h(x) be the hazard rate belonging to F .

Since $$h_{1,n,\tilde{m},k}(x) = c_0\, h(x) = (k + n - 1 + M_1)\, h(x) ,$$

we obtain the IFR property of an arbitrary g OS, if the underlying distribution has this property. By this, we find a generalization of the results for o OS's (see Barlow, Proschan

1975, p 108; cf. Takahasi 1988, p 4135) and for records (Gupta, Kirmani 1988, Kochar 1990). (For any given n we consider $r = 1$ which yields the assertion via the multiplicative structure shown above.)

With the same arguments as in Theorem 1.5. we obtain

THEOREM 1.7. Let $r \le n-1$, $m_1 = \dots = m_r = m$. Then we find :
If $X(r+1,n,\tilde{m},k)$ possesses the DFR property, so does $X(r,n,\tilde{m},k)$.

Nagaraja (1990) shows the transmission of the IFR property from the r–th o OS to the r–th o OS in a sample of smaller size. This result can also be generalized to g OS's. In order to have a simple multiplicative relationship between the constants $c_{r-j-1}(n+1)$ and $c_{r-j-1}(n)$ which depend on $\tilde{m}$, we consider the special case of identical m_i ($m_1 = \dots = m_n$). To prepare for the proof, we give the following lemma (cf. Nagaraja 1990, p 309).

LEMMA 1.8. The function η_2 with

$$\eta_2(t) = \frac{\sum\limits_{j=0}^{r-1} \frac{1}{j!\, c_{r-j-1}(n+1)}\, t^j}{\sum\limits_{j=0}^{r-1} \frac{1}{j!\, c_{r-j-1}(n)}\, t^j}, \quad r \le n,$$

$\left\{ \begin{array}{l} \text{increases} \\ \text{decreases} \end{array} \right.$ with respect to $t \ge 0$, if $\left\{ \begin{array}{l} m \ge -1 \\ m < -1 \end{array} \right.$, and has range

$$\left[\frac{k + (n-r)(m+1)}{k + n(m+1)}, \frac{k + (n-1)(m+1)}{k + n(m+1)} \right).$$

PROOF η_2 is constant, if $r = 1$ or $m = -1$. Let $r \ge 2$.
Using Lemma 1.2. with n and r replaced by $r-1$, we have :

$(m+1)\, \eta_1^{(r-1,n)}(t) + k + (n-r)(m+1)$

$$= (m+1) \frac{\sum\limits_{j=0}^{r-2} \frac{1}{j!\, c_{r-j-2}}\, t^{j+1}}{\sum\limits_{j=0}^{r-1} \frac{1}{j!\, c_{r-j-1}}\, t^j} + k + (n-r)(m+1)$$

$$= \frac{1}{\sum\limits_{j=0}^{r-1} \frac{1}{j!\, c_{r-j-1}} t^j} \left((m+1) \sum_{j=0}^{r-1} \frac{1}{(j-1)!\, c_{r-j-1}} t^j + (k+(n-r)(m+1)) \sum_{j=0}^{r-1} \frac{1}{j!\, c_{r-j-1}} t^j \right)$$

$$= \frac{1}{\sum\limits_{j=0}^{r-1} \frac{1}{j!\, c_{r-j-1}} t^j} \left(\frac{1}{c_{r-2}} + (k+n(m+1)) \sum_{j=1}^{r-1} \frac{1}{j!\, c_{r-j-1}(n+1)} t^j \right)$$

$$= (k + n(m + 1))\, \eta_2(t)\,.$$

THEOREM 1.9. Let $r \leq n$, $m_1 = \ldots = m_n = m$. Then we find in the case $m \geq -1$ $(m < -1)$ (respectively) :

If $X(r,n+1,m,k)$ possesses the IFR (DFR) property, so does $X(r,n,m,k)$.

PROOF Let $t = t(x) = g_m(F(x))$ as in the proof of Theorem 1.5.

Then the assertion follows from $\frac{h_{r,n,m,k}(x)}{h_{r,n+1,m,k}(x)} = \eta_2(t)$ together with Lemma 1.8.

The same arguments lead to

THEOREM 1.10. Let $r \leq n$, $m_1 = \ldots = m_n = m$. Then we find in the case $m \geq -1$ $(m < -1)$ (respectively) :

If $X(r,n,m,k)$ possesses the DFR (IFR) property, so does $X(r,n+1,m,k)$.

In the case of records, i.e. $m = -1$, the assertions are trivial, since the marginal density function of $X(r,n,-1,k)$ does not depend on n ; hence, the g OS's $X(r,n+1,-1,k)$ and $X(r,n,-1,k)$ have identical distributions.

In view of identical distributions of

$$X(r,n+1,m,k) \text{ and } X(r,n,m,k+m+1)$$

(see I.3.1.6.), the results may be interpreted as relations between the g OS's $X(r,n,m,k)$ and $X(r,n,m,k+m+1)$; i.e., we find a transmission result with respect to the parameter k.

Finally, we state a result for g OS's by analogy with the transmission of the IFR property from the o OS $X_{r,n}$ to $X_{r+1,n+1}$ which is shown in Nagaraja (1990). In the case of records, the following assertions (X(r+1,n+1,m,k) and X(r+1,n,m,k) are identically distributed) coincide with the ones given in Kochar (1990). In the introductory remarks we refer to another result in which the choice of r and n is restricted; we do not give a generalization here.

LEMMA 1.11. The function η_3 with

$$\eta_3(t) = \frac{\sum_{j=0}^{r-1} \frac{1}{j!\, c_{r-j-1}(n)}\, t^{j+1}}{\sum_{j=0}^{r} \frac{1}{j!\, c_{r-j}(n+1)}\, t^{j}}$$

increases strictly with respect to $t \geq 0$ if $r \leq n-1$ or if $r = n$, $k \geq m+1$,

and has range $$[0\,,\, r\,\frac{k + n(m+1)}{k + (n-1)(m+1)}\,).$$

PROOF The proof is obtained using Lemma 1.2. (1.3. ii)) (cf. Nagaraja 1990, p 310).

We have : $$\frac{\eta(t)}{k + (n-r-1)(m+1) + (m+1)\,\eta(t)}$$

$$= \frac{\sum_{j=0}^{r-1} \frac{1}{j!\, c_{r-j-1}(n)}\, t^{j+1}}{\sum_{j=1}^{r} \frac{1}{j!\, c_{r-j}(n)}\, t^{j}} \left(k+(n-r-1)(m+1) + (m+1)\, \frac{\sum_{j=0}^{r-1} \frac{1}{j!\, c_{r-j-1}(n)}\, t^{j+1}}{\sum_{j=1}^{r} \frac{1}{j!\, c_{r-j}(n)}\, t^{j}}\right)^{-1}$$

$$= \frac{1}{k + n(m+1)}\,\tilde{\eta}_3(t)\,,\ \text{say, with}\ \ \tilde{\eta}_3(t) = \frac{\sum_{j=0}^{r-1} \frac{1}{j!\, c_{r-j-1}(n)}\, t^{j+1}}{\sum_{j=1}^{r} \frac{1}{j!\, c_{r-j}(n+1)}\, t^{j}}.$$

Let $r \leq n-1$:

Since η increases with respect to t (see 1.3. ii)) and the function on the l.h.s. increases with respect to η, we see that $\tilde{\eta}_3$ increases with respect to t.

In the case $r = n$, we observe that $\tilde{\eta}_3$ $\left\{\begin{array}{l}\text{increases}\\ \text{decreases}\end{array}\right.$ with respect to t, if $\left\{\begin{array}{l} k \geq m+1 \\ k < m+1 \end{array}\right.$.

In those cases in which $\tilde{\eta}_3$ is a monotonically increasing function with respect to t, we obtain monotonicity of η_3 by adding the constant, nonnegative term $\frac{1}{c_r(n+1)}$ to the denominator.

THEOREM 1.12. Let $r \leq n$, $m_1 = \dots = m_n = m$, and $k \geq m+1$, if $r = n$. Then we find: If $X(r,n,m,k)$ possesses the IFR property, so does $X(r+1,n+1,m,k)$.

PROOF Again, let $t = t(x) = g_m(F(x))$.
Together with Lemma 1.11. the assertion follows from $\frac{h_{r+1,n+1,m,k}(x)}{h_{r,n,m,k}(x)} = \frac{1}{r}\,\eta_3(t)$.

The same argument leads to

THEOREM 1.13. Let $r \leq n$, $m_1 = \dots = m_n = m$, and $k \geq m+1$, if $r = n$. Then we find: If $X(r+1,n+1,m,k)$ possesses the DFR property, so does $X(r,n,m,k)$.

There are further results on the transmission of the DFR property. It is shown by Takahasi (1988) and Kochar (1990) that there does not exist a distribution such that all order statistics or record values have DFR distributions. These results are contained in

THEOREM 1.14. For every $m \geq -1$ and for every $k \geq 1$ there exist $r = r(F,m,k)$ and $n = n(F,m,k)$, $r \leq n$, such that the distribution of $X(r,n,m,k)$ does not possess the DFR property.

PROOF is obtained along the lines of Takahasi (1988).
Choose $x_1 < x_2$ satisfying $f(x_1) \neq 0$, $f(x_2) \neq 0$ and $0 < F(x_1) < F(x_2) < 1$.
Then it can be shown that there exists an integer n_0 such that

$$h_{n_0,n_0,m,k}(x_1) < h_{n_0,n_0,m,k}(x_2)\,;$$

i.e. $h_{n_0,n_0,m,k}$ is not decreasing.

If $m \geq -1$, then the g OS's $X(r,n,m,k)$ are well defined for all $n \geq 1$. In contrast to this, in the case $m < -1$ the integer n has to be chosen less or equal some upper bound $n^* = n^*(m,k)$, since $\gamma_r \geq 1$ has to be ensured for all $r = 1,\dots,n$. Now, fixing some $m < -1$ and $k \geq 1$, and by this the corresponding n^*, it is possible that all records $X(r,n,m,k)$, $n \leq n^*$, possess the DFR property:

If $X(n^*,n^*,m,k)$ has a DFR distribution, then Theorems 1.7. and 1.9. together imply the DFR property of the distributions of all $X(r,n,m,k)$, $1 \leq r \leq n \leq n^*$.

However, the DFR property of some distribution function F is transmitted to the distribution functions of the spacings

$$X(r,n,m,k) - X(r-1,n,m,k)\ ,\ r \geq 2\ ,$$

which is shown by Gupta, Kirmani (1988) in the case of ordinary record values.

Theorem 1.15. If the underlying distribution function F of the g OS's $X(r,n,m,k)$, $1 \leq r \leq n$, possesses the DFR property, then all spacings $X(r,n,m,k) - X(r-1,n,m,k)$, $r \geq 2$, have DFR distributions.

Proof Using the Markovian structure and putting

$$F_s(t) = 1 - \left(\frac{1 - F(t+s)}{1 - F(s)}\right)^{\gamma_r}, \ t > 0\,,\ s > 0\,,$$

we have $$P(\,X(r,n,m,k) - X(r-1,n,m,k) \leq t\,) = \int_0^\infty F_s(t)\, d\,G(s)$$

where G is the distribution function of $X(r-1,n,m,k)$.

Since the failure rate of F_s is given by $\gamma_r \cdot \frac{f(t+s)}{1 - F(t+s)}$, the DFR property of F implies the DFR property of F_s.

Thus, the distribution of a spacing being a mixture of DFR distributions, the assertion follows (see Ross 1983, p 254).

Considering the diagram of the transmission of the IFR property, e.g., there is a certain gap. That is, we do not have a description, a characterization of that region where the IFR property is transmitted to.

REMARK Concerning the IFR property we show above that, without any restriction of the parameters, we have transmission of this property along the line given by $(r,n) \longmapsto (r+1,n+1)$ and on its r.h.s. (see the diagram). Under a certain restriction of the parameters, Nagaraja (1990) obtains a separating line with a smaller derivative.
Thus we may ask whether any (useful) derivative of this separating line, combined with corresponding transmission rules, can be obtained restricting the parameters r and n appropriately.

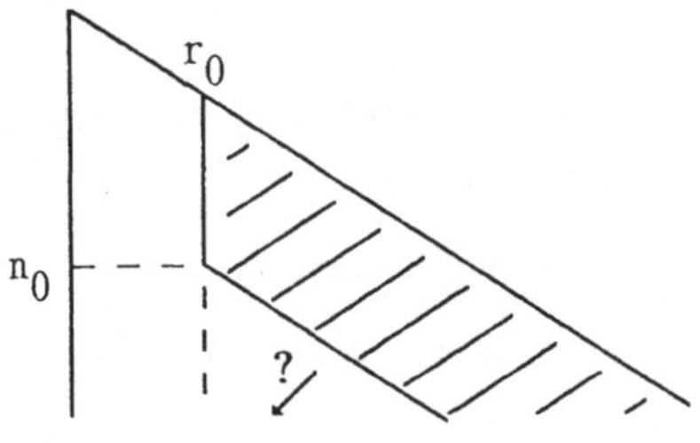

The above results (1.5. – 1.13.) on IFR and DFR properties can be extended to IFRA and NBU properties and to DFRA and NWU properties, respectively (cf. Nagaraja 1990). In the case of o OS's, Nagaraja applies a result of Block, Borges, Savits (1985, p 373).
Here, the extension can also be seen directly. In the Lemmas 1.2., 1.8. and 1.11. we prove monotonicity of certain ratios of failure rates. So we can use the following argument :

REMARK 1.16. Let the failure rates

$$\lambda_i = \frac{f_i(t)}{1 - F_i(t)}, \; i = 1,2,$$

be given satisfying $\lambda_1(t)/\lambda_2(t) = p(t)$ with p increasing with respect to t.

i) Transmission of the IFRA property :

Let $$\mu_i(t) = -\tfrac{1}{t} \log(1 - F_i(t)) .$$

Then we have :

$\mu_1(t) = p(\xi_t)\, \mu_2(t)$ with $0 \le \xi_t \le t$ and ξ_t increasing with respect to t .

Thus we may adopt the argument from the IFR case.

ii) Transmission of the NBU property :

We have

$$1 - F_1(x + y) \le (1 - F_1(x))\,(1 - F_1(y))$$
$$\Leftrightarrow\ (x + y)\,\mu_1(x + y) \ge x\,\mu_1(x) + y\,\mu_1(y)$$
$$\Leftrightarrow\ p(\xi_{x+y})\log(1 - F_2(x+y)) \le p(\xi_x)\log(1 - F_2(x)) + p(\xi_y)\log(1 - F_2(y))\,.$$

Thus if the distribution function F_2 possesses the NBU property, we obtain

$$\log(1 - F_2(x+y)) \le \log(1 - F_2(x)) + \log(1 - F_2(y))$$

for all $x, y \ge 0$.

Hence F_1 possesses the NBU property .

Analogously, we may transfer the DFR results to corresponding DFRA and NWU transmission properties.
Concerning the DMRL property, we refer to the article of Nagaraja (1990, p 312/3).

2. Partial Ordering of Generalized Order Statistics

In the literature, we find many proposals of stochastic order relations between r.v.'s or distributions and there is a variety of related results. For a survey on partial orderings we refer to, e.g., Shaked (1982), Ross (1983), Bagai, Kochar (1986), Bartoszewicz (1987) and to articles by Dykstra (1985), Oja (1985), Shaked (1985) and Whitt (1988) in the Encyclopedia of Statistical Sciences.

Let throughout X and Y be r.v.'s with distribution functions F and G respectively, satisfying

$$F^{-1}(0)\,,\ G^{-1}(0) \ge 0\,;$$

i.e., their supports are contained in the positive real line.
Whenever necessary, it is assumed that F and G are absolutely continuous with corresponding density functions f and g.
We now introduce some partial orderings between r.v.'s (or distributions).
Stochastic ordering, failure rate ordering and likelihood ratio ordering are well known and can be found in Ross (1983), e.g.

2.1. Stochastic Ordering :

X is said to be stochastically smaller than Y,

written $$X \overset{st.}{\leq} Y,$$

iff $$1 - F(x) \leq 1 - G(x) \quad \text{for all } x \geq 0 .$$

2.2. Failure Rate Ordering :

X has a smaller failure rate than Y,

written $$X \overset{f.r.}{\leq} Y,$$

iff $$\frac{f(x)}{1 - F(x)} \leq \frac{g(x)}{1 - G(x)} \quad \text{for all } x \geq 0 .$$

2.3. Likelihood Ratio Ordering :

X is said to be smaller than Y in the sense of likelihood ratio,

written $$X \overset{l.r.}{\leq} Y,$$

iff $$\frac{f(x)}{g(x)} \geq \frac{f(y)}{g(y)} \quad \text{for all } 0 \leq x \leq y .$$

We observe that likelihood ratio ordering implies failure rate ordering ($X \overset{l.r.}{\leq} Y \Rightarrow X \overset{f.r.}{\geq} Y$; e.g. Ross 1983) which itself implies stochastic ordering ($X \overset{f.r.}{\geq} Y$ and $G(0) \leq F(0) \Rightarrow X \overset{st.}{\leq} Y$; cf. Gupta, Kirmani 1987, Cor. 1).

In Doksum (1969) the tail–ordering of distributions is introduced (cf. Deshpande, Kochar

1983). Bagai, Kochar (1986) point out the relations between tail–ordering and failure rate ordering depending on an increasing or decreasing failure rate. Namely, if F or G is IFR (DFR), then tail–ordering is stronger (weaker) than failure rate ordering.

2.4. Tail – Ordering :

X is said to be tail–ordered with respect to Y,

written $X \overset{t}{\leq} Y$,

iff $G^{-1}(F(x)) - x$ is non–decreasing in x , $0 \leq x < F^{-1}(1)$.

An intuitive approach to some partial ordering is to compare quantile differences; see Lewis, Thompson (1981) and Shaked (1982, 1985).

2.5. Dispersive Ordering :

X is said to be dispersed with respect to Y,

written $X \overset{\text{disp.}}{\leq} Y$,

iff $F^{-1}(\beta) - F^{-1}(\alpha) \leq G^{-1}(\beta) - G^{-1}(\alpha)$ for all $0 < \alpha < \beta < 1$.

If X and Y are ordered in dispersion, i.e. $X \overset{\text{disp.}}{\leq} Y$, then we obtain $\text{Var } X \leq \text{Var } Y$, e.g. (cf. Shaked 1982).

However, tail–ordering and dispersive ordering are essentially the same as pointed out by Deshpande, Kochar (1983).

To be precise, if the distribution function F is continuous and strictly increasing, then the dispersive ordering implies the tail–ordering. Continuity of F is sufficient to prove the converse implication.

In the literature there are some articles investigating partial ordering in the case of order statistics and record values; e.g. Bartoszewicz (1985), Gupta, Kirmani (1988, 1990), Kochar (1990).

Analogous properties can also be obtained for g OS′s based on an absolutely continuous distribution function F.

Moreover, we refer to Kim, Proschan, Sethuraman (1988) and Barbour, Lindvall, Rogers (1991) for stochastic ordering of order statistics.

Kochar (1990) shows that ordinary record values are partially ordered in the sense of likelihood ratio. This result is contained in

THEOREM 2.6. Let $X(r_1,n_1,m,k_1)$ and $X(r_2,n_2,m,k_2)$ be g OS's based on F,

let $\quad A = k_1 - k_2 + (n_1 - n_2 - (r_1 - r_2))(m + 1)$

and $\quad B = k_1 - k_2 + (n_1 - n_2)(m + 1)$.

If the parameters satisfy the conditions

(*) $$r_1 \geq r_2 \quad \text{and} \quad \min(A,B) \leq 0\,,$$

then we find $$X(r_1,n_1,m,k_1) \overset{l.r.}{\geq} X(r_2,n_2,m,k_2)\,.$$

PROOF is obtained determining r_1, r_2 and A such that $(1-x)^A\, g_m^{\,r_1-r_2}(x)$ increases.

EXAMPLES

i) In the case of o OS's ($m = 0$, $k_1 = k_2 = 1$), condition (*) simplifies to

$$r_1 \geq r_2 \quad \text{and} \quad n_1 - n_2 \leq r_1 - r_2\,.$$

ii) In the case of k–th record values ($m = -1$), condition (*) is simply

$$r_1 \geq r_2 \quad \text{and} \quad k_1 \leq k_2\,.$$

(cf. Kochar 1990 for $k_1 = k_2 = 1$).

Moreover, in the situation of Theorem 2.6. we obtain that the corresponding hazard rates are ordered : $$h_{r_1,n_1,m,k_1}(x) \leq h_{r_2,n_2,m,k_2}(x) \quad \text{for all } x \geq 0\,,$$

since likelihood ratio ordering is stronger than failure rate ordering.

Finally, we state a result on stochastic ordering of differences of g OS's assuming tail–ordering of the underlying distribution functions. As particular cases, we obtain the results of Bartoszewicz (1985) for o OS's ($m = 0$) and of Kochar (1990) for record values ($m = -1$).

THEOREM 2.7. Let $X(i,n,\tilde{m},k)$, $Y(i,n,\tilde{m},k)$, $1 \leq i \leq n$, be g OS's based on F and G respectively, let F be continuous and let $X \overset{t}{\leq} Y$, $X \sim F$, $Y \sim G$.

Then for $1 \leq r \leq s \leq n$, $m_1 = \dots = m_{s-1} = m$, we have

$$X(s,n,\tilde{m},k) - X(r,n,\tilde{m},k) \overset{st.}{\leq} Y(s,n,\tilde{m},k) - Y(r,n,\tilde{m},k) .$$

PROOF Since $F^{X(i,n,\tilde{m},k)}(x) = \Phi_{i,n}(F(x))$, $m_1 = \dots = m_{i-1} = m$ (cf. I.3.2.6.), we have

$$P (G^{-1} F (X(i,n,\tilde{m},k)) \leq x) = P (F (X(i,n,\tilde{m},k)) \leq G(x))$$

$$= P (X(i,n,\tilde{m},k) \leq F^{-1}(G(x) + 0)) = \Phi_{i,n}(G(x)) = P (Y(i,n,\tilde{m},k) \leq x) .$$

The assumption $X \overset{t}{\leq} Y$ and $X(r,n,\tilde{m},k) \leq X(s,n,\tilde{m},k)$ together imply

$$G^{-1} F (X(r,n,\tilde{m},k)) - X(r,n,\tilde{m},k) \leq G^{-1} F (X(s,n,\tilde{m},k)) - X(s,n,\tilde{m},k)$$

and hence the assertion.

Appendix

1. Integration Formulae

In the proofs of the assertions in I.3.2.11., II.1.1.3. and IV.1.2.1., we use integration formulae which are cited in Lemma A.1.1. and then stated in terms of an integral involving the function g_m (see I.2.10.).

LEMMA 1.1.

i) With constants μ, ν, $m+1 > 0$ we have :

$$\int_0^1 x^{\mu-1}\,(1-x^{m+1})^{\nu-1}\,dx = \frac{1}{m+1}\,B\left(\frac{\mu}{m+1},\nu\right),\quad B(a,b) = \frac{\Gamma(a)\,\Gamma(b)}{\Gamma(a+b)}$$

(Gröbner, Hofreiter 2, 1961, p 18, Gradstein, Ryshik 1, 1981, p 347).

ii) With constants μ, $\nu > 0$ we have :

$$\int_0^1 x^{\mu-1}\,\left(\log\frac{1}{x}\right)^{\nu-1}\,dx = \mu^{-\nu}\,\Gamma(\nu)$$

(Gröbner, Hofreiter 2, 1961, p 74, Gradstein, Ryshik 1, 1981, p 597).

COROLLARY 1.2. Let the function g_m as in I.2.10. and constants μ, $\nu > 0$ be given with

$$\mu + (\nu-1)(m+1) > 0\,.$$

Then we find :

$$\int_0^1 (1-t)^{\mu-1}\,g_m^{\nu-1}(t)\,dt = \begin{cases} (m+1)^{-\nu}\;B\left(\frac{\mu}{m+1},\nu\right) & ,\ m > -1 \\ \mu^{-\nu}\;\Gamma(\nu) & ,\ m = -1 \\ (-m-1)^{-\nu}\;B\left(-\frac{\mu}{m+1}-\nu+1,\nu\right) & ,\ m < -1 \end{cases}$$

$$= \Gamma(\nu)\left(\prod_{i=0}^{\nu-1}(\mu + i(m+1))\right)^{-1}.$$

2. Combinatorial Identities

In this section we gather together some combinatorial identities which turn out to be useful in the analysis of o OS's and g OS's (see I.3.1.2., Sections I.3.2., V.1.4.).

LEMMA 2.1. If $1 \le r \le n$ and $x \in \mathbb{R}$ we have :

$$\sum_{j=0}^{r-1} \binom{n-r+j}{j} x^j = \sum_{j=0}^{r-1} \binom{n}{j} x^j (1-x)^{r-j-1} .$$

PROOF The assertion follows by induction on r starting with $r = 1$:

$$\sum_{j=0}^{r} \binom{n}{j} x^j (1-x)^{r-j} = (1-x) \sum_{j=0}^{r-1} \binom{n}{j} x^j (1-x)^{r-j-1} + \binom{n}{r} x^r$$

$$= (1-x) \sum_{j=0}^{r-1} \binom{n-r+j}{j} x^j + \binom{n}{r} x^r$$

$$= 1 + \sum_{j=1}^{r} \left(\binom{n-r+j}{j} - \binom{n-r+j-1}{j-1} \right) x^j = \sum_{j=0}^{r} \binom{n-r-1+j}{j} x^j .$$

LEMMA 2.2. Let $r \in \mathbb{N}_0$, $a \neq 0$ and $b \in \mathbb{R}$ be given satisfying $aj + b \neq 0$ for all $0 \le j \le r$. Then we observe :

$$\sum_{j=0}^{r} \frac{1}{j!\ (r-j)!} (-1)^j \frac{1}{aj+b} x^{aj+b} = x^b a^r \sum_{j=0}^{r} \frac{1}{j!} \left(\prod_{\nu=j}^{r} (a\nu + b) \right)^{-1} \left(\tfrac{1}{a}(1-x^a)\right)^j .$$

PROOF The assertion follows by induction on r starting with $r = 0$:

$$x^b a^{r+1} \sum_{j=0}^{r+1} \frac{1}{j!} \left(\prod_{\nu=j}^{r+1} (a\nu + b) \right)^{-1} \left(\tfrac{1}{a}(1-x^a)\right)^j$$

$$= \frac{a}{a(r+1)+b} x^b a^r \sum_{j=0}^{r} \frac{1}{j!} \left(\prod_{\nu=j}^{r} (a\nu + b) \right)^{-1} \left(\tfrac{1}{a}(1-x^a)\right)^j$$

$$+ x^b a^{r+1} \frac{1}{(r+1)!} \frac{a}{a(r+1)+b} \left(\tfrac{1}{a}(1-x^a)\right)^{r+1}$$

$$= \frac{a}{a(r+1)+b} \sum_{j=0}^{r} \frac{1}{j!\ (r-j)!} (-1)^j \frac{1}{aj+b} x^{aj+b} + x^b \frac{1}{(r+1)!} \frac{a}{a(r+1)+b} \sum_{j=0}^{r+1} \binom{r+1}{j} (-1)^j x^{aj}$$

$$= \sum_{j=0}^{r+1} \frac{1}{j!\ (r+1-j)!} (-1)^j \frac{1}{aj+b} x^{aj+b} .$$

COROLLARY 2.3. Let $r \in \mathbb{N}_0$.

i) $$\sum_{j=0}^{r} (-1)^j \binom{r}{j} \frac{1}{aj+b} = a^r r! \left(\prod_{\nu=0}^{r} (a\nu + b) \right)^{-1}, \text{ if } aj+b \neq 0 \text{ for all } 0 \leq j \leq r ;$$

ii) $$\sum_{j=0}^{r} (-1)^j \binom{r}{j} \frac{1}{2j+1} = 2^r r! \left(\prod_{\nu=0}^{r} (2\nu + 1) \right)^{-1} = \frac{(2r)!!}{(2r+1)!!} .$$

LEMMA 2.4. If $1 \leq r \leq n$ and $x \in \mathbb{R}$ we find :

$$\frac{n!}{(n-r)!} \sum_{j=0}^{r-1} (-1)^j \frac{1}{j!\ (r-1-j)!} \frac{(1-x)^{n-r+j+1}}{n-r+j+1}$$

$$\left(= \frac{n!}{(n-r)!} \sum_{j=0}^{r-1} (-1)^{r-1-j} \frac{1}{j!\ (r-1-j)!} \frac{(1-x)^{n-j}}{n-j} \right)$$

$$= \sum_{j=0}^{r-1} \binom{n}{j} x^j (1-x)^{n-j} .$$

PROOF Using the Lemmas 2.2. and 2.1. we observe

$$\frac{n!}{(n-r)!} \sum_{j=0}^{r-1} (-1)^j \frac{1}{j!\ (r-1-j)!} \frac{(1-x)^{n-r+j+1}}{n-r+j+1} = (1-x)^{n-r+1} \sum_{j=0}^{r-1} \binom{n-r+j}{j} x^j$$

$$= (1-x)^{n-r+1} \sum_{j=0}^{r-1} \binom{n}{j} x^j (1-x)^{r-j-1} = \sum_{j=0}^{r-1} \binom{n}{j} x^j (1-x)^{n-j} .$$

A modification of the assertion of Lemma 2.1. may be obtained by

LEMMA 2.5. Let $\mu \in \mathbb{N}_0$, $r, n \in \mathbb{N}$ be given with $\mu + 1 \leq r \leq n$. Then we have :

$$\sum_{j=0}^{\mu} (-1)^j \binom{n}{j} \binom{r-j-1}{\mu-j} = (-1)^{\mu} \binom{n-r+\mu}{\mu} .$$

PROOF The assertion follows by induction on n using 2.3.

Symbols

$\mathbb{N} = \{1, 2, \dots\}$	positive integers, natural numbers	
$\mathbb{N}_0 = \{0, 1, 2, \dots\}$		
$\mathbb{Z} = \{\dots, -2, -1, 0, 1, 2, \dots\}$		
$\mathbb{R}^n$	n–dimensional Euclidean space	
$\mathbb{R} = \mathbb{R}^1$		
iff	if and only if	
$X, X_1, \dots, X_n$	random variables	
r.v.'s	random variables	
iid	independent and identically distributed	
$F, F_1, \dots, F_n$	distribution functions	
$f, f_1, \dots, f_n$	density functions	
o OS	ordinary order statistic	Introduction
g OS	generalized order statistic	Introduction
OS's	order statistics	Introduction
$X_{r,n}$	order statistic	I.1.1.
F^{-1}	pseudo–inverse	I.1.1.2.
$X_{r,\alpha}$	order statistic with non–integral sample size	I.1.2.
$X_*^{(r)}$	sequential order statistic	I.1.3.
$\alpha_r > 0$	parameter	I.1.3.4.
$X_{L(r)}$	record value	I.1.4.
$X_{L^{(k)}(r)}$	k–th record value	I.1.5.
$X_{\Delta_r}^{(r)}$	Pfeifer's record value	I.1.6.
$\beta_r > 0$	parameter	I.1.6.8.

$X^{(r)}_{\Delta_r, k_r}$	k_r–record	I.1.7.
$X^{(r)}_{*, k_r}$		I.1.8.
$r, n \in \mathbb{N}, \ r \leq n$	parameters	I.2.1.
$k \geq 1$	parameter	I.2.1.
$m_1, \dots, m_{n-1} \in \mathbb{R}$	parameters	I.2.1.
$M_r = \sum_{j=r}^{n-1} m_j$		I.2.1.
$\sum_{i=\nu}^{\mu} \dots = 0, \quad \text{if } \mu < \nu$		
$\gamma_r = k + n - r + M_r \geq 1$		I.2.1.
$\tilde{m} = (m_1, \dots, m_{n-1})$	vector of parameters	I.2.1.
$U(r,n,\tilde{m},k)$	uniform g OS	I.2.1.
$U(r,n,m,k)$		I.2.1.
$X(r,n,\tilde{m},k)$	g OS	I.2.3.
$X(r,n,m,k)$		I.2.3.
$c_{r-1} = \prod_{i=1}^{r} \gamma_i$		I.2.10.
$\prod_{i=\nu}^{\mu} \dots = 1, \quad \text{if } \mu < \nu$		
$h_m(x) = \begin{cases} -\frac{1}{m+1}(1-x)^{m+1}, & m \neq -1 \\ \log\left(\frac{1}{1-x}\right), & m = -1 \end{cases}, \quad x \in [0,1)$		I.2.10.
$g_m(x) = h_m(x) - h_m(0) = \begin{cases} \frac{1}{m+1}(1-(1-x)^{m+1}), & m \neq -1 \\ \log\left(\frac{1}{1-x}\right), & m = -1 \end{cases}, \quad x \in [0,1)$		I.2.10.
$\varphi_{r,n}(x) = \frac{c_{r-1}}{(r-1)!}(1-x)^{\gamma_r - 1} g_m^{r-1}(x), \ x \in (0,1)$		I.3.2.1.
$\Phi_{r,n}(x) = \int_0^x \varphi_{r,n}(t)\, dt$		I.3.2.4.

References

Ahsanullah, M. (1987)
Record statistics and the exponential distribution
Pak. J. Statistics 3A, 17–40

Ahsanullah, M. (1989)
Estimation of the parameters of a power function distribution by record values
Pak. J. Statistics 5A, 189–194

Ahsanullah, M., Houchens, R.L. (1989)
A note on record values from a Pareto distribution
Pak. J. Statistics 5A, 51–57

Ahsanullah, M., Rahman, M. (1972)
A characterization of the exponential distribution
J. Appl. Prob. 9, 457–461

Aly, M.A.H. (1988)
Some contributions to characterization theory with applications in stochastic processes
Doctoral dissertation, University of Sheffield

Arnold, B.C. (1985)
p–Norm bounds on the expectation of the maximum of a possibly dependent sample
J. Mult. Analysis 17, 316–332

Arnold, B.C. (1988)
Bounds on the expected maximum
Commun. Statist. – Theory Meth. 17, 2135–2150

Arnold, B.C., Balakrishnan, N. (1989)
Relations, Bounds and Approximations for Order Statistics
Springer, Berlin

Arnold, B.C., Balakrishnan, N., Nagaraja, H.N. (1992)
A First Course in Order Statistics
Wiley, New York

Arnold, B.C., Becker, A., Gather, U., Zahedi, H. (1984)
On the Markov property of order statistics
J. Stat. Plan. Inf. 9, 147–154

Arnold, B.C., Meeden, G. (1975)
Characterization of distributions by sets of moments of order statistics
Ann. Statist. 3, 754–758

Ascher, H.E., Feingold, H. (1984)
Repairable Systems Reliability
Dekker, New York

Azlarov, T.A., Volodin, N.A. (Stein, M. trans.) (1986)
Characterization Problems Associated with the Exponential Distribution
Springer, New York

Bagai, I., Kochar, S.C. (1986)
On tail–ordering and comparison of failure rates
Commun. Statist. – Theory Meth. 15, 1377–1388

Balakrishnan, N. (1987)
A note on moments of order statistics from exchangeable variates
Commun. Statist. – Theory Meth. 16, 855–861

Balakrishnan, N. (1993)
A simple application of binomial–negative binomial relationship in the derivation of sharp bounds for moments of order statistics based on greatest convex minorants
Statistics & Probability Letters 18, 301–305

Balakrishnan, N., Balasubramanian, K. (1993)
Equivalence of Hartley–David–Gumbel and Papathanasiou bounds and some further remarks
Statistics & Probability Letters 16, 39–41

Balakrishnan, N., Bendre, S.M., Malik, H.J. (1992)
General relations and identities for order statistics from non–independent non–identical variables
Ann. Inst. Statist. Math. 44, 177–183

Balakrishnan, N., Cohen, A.C. (1991)
Order Statistics and Inference: Estimation Methods
Academic Press, Boston

Balakrishnan, N., Joshi, P.C. (1982)
Moments of order statistics from doubly truncated Pareto distribution
J. Indian Statist. Ass. 20, 109–117

Balakrishnan, N., Malik, H.J., Ahmed, S.E. (1988)
Recurrence relations and identities for moments of order statistics, II: Specific continuous distributions
Commun. Statist. – Theory Meth. 17, 2657–2694

Balasubramanian, K., Bapat, R.B. (1991)
Identities for order statistics and a theorem of Renyi
Statistics & Probability Letters 12, 141–143

Balasubramanian, K., Balakrishnan, N. (1993)
Duality principle in order statistics
J. R. Statist. Soc. B 55, 687–691

Barbour, A.D., Lindvall, T., Rogers, L.C.G. (1991)
Stochastic ordering of order statistics
J. Appl. Prob. 28, 278–286

Barlow, R.E., Proschan, F. (1965)
Mathematical Theory of Reliability
Wiley, New York

Barlow, R.E., Proschan, F. (1975)
Statistical Theory of Reliability and Life Testing, Probability Models
Holt–Rinehart and Winston, New York

Barlow, R.E., Proschan, F. (1981)
Statistical Theory of Reliability and Life Testing
To Begin With, Silver Spring

Barnett, V.D. (1966)
Order statistics estimators of the location of the Cauchy distribution
J. Am. Statist. Ass. 61, 1205–1218
Correction 63, 383–385

Barnett, V. (1988)
Outliers and order statistics
Commun. Statist. – Theory Meth. 17, 2109–2118

Barnett, V., Lewis, T. (1978, 1994 3rd ed.)
Outliers in Statistical Data
Wiley, New York

Bartoszewicz, J. (1985)
Moment inequalities for order statistics from ordered families of distributions
Metrika 32, 383–389

Bartoszewicz, J. (1987)
A note on dispersive ordering defined by hazard functions
Statistics & Probability Letters 6, 13–16

Basu, A.D. (1988)
Probabilistic Reliability
in: Kotz, S., Johnson, N.L. (eds.), Encyclopedia of Statistical Sciences, Vol. 8
Wiley, New York

Beckenbach, E.F., Bellman, R. (1961)
Inequalities
Springer, Berlin

Bhattacharya, P.K. (1974)
Convergence of sample paths of normalized sums of induced order statistics
Ann. Statist. 2, 1034–1039

Bhattacharya, P.K. (1984)
Induced order statistics: Theory and applications
in: Krishnaiah, P.R., Sen, P.K. (eds.), Handbook of Statistics, Vol. 4, 383–403
North Holland, Amsterdam

Block, H.W., Borges, W.S., Savits, T.H. (1985)
Age–dependent minimal repair
J. Appl. Prob. 22, 370–385

Blom, G. (1958)
Statistical Estimates and Transformed Beta–Variables
Almqvist & Wiksell, Stockholm

Boas, R.P. (1954)
Entire Functions
Academic Press, New York

Cacoullos, T. (1989)
Exercises in Probability
Springer, New York

Castillo, E. (1988)
Extreme Value Theory in Engineering
Academic Press, Boston

Chan, L.K. (1967)
On a characterization of distributions by expected values of extreme order statistics
American Math. Monthly 74, 950–951

Chandler, K.N. (1952)
The distribution and frequency of record values
J. R. Statist. Soc. B 14, 220–228

Choudhury, J., Serfling, R.J. (1988)
Generalized order statistics, Bahadur representations, and sequential nonparametric fixed width confidence intervals
J. Stat. Plan. Inf. 19, 269–282

Cole, R.H. (1951)
Relations between moments of order statistics
Ann. Math. Statist. 22, 308–310

Crawford, G.B. (1966)
Characterization of geometric and exponential distributions
Ann. Math. Statist. 37, 1790–1795

David, H.A. (1973)
Concomitants of order statistics
Bull. Inst. Internat. Statist. 45, 295–300

David, H.A. (1981)
Order Statistics
2 nd ed., Wiley, New York

David, H.A. (1988)
General bounds and inequalities in order statistics
Commun. Statist. – Theory Meth. 17, 2119–2134

David, H.A. (1993a)
A note on order statistics for dependent variates
The American Statistician 47, 198–199

David, H.A. (1993b)
Concomitants of order statistics: Review and recent developments
in: Hoppe, F.M. (ed.), Multiple Comparisons, Selection, and Applications in Biometry, 507–518
Dekker, New York

David, H.A., Galambos, J. (1974)
The asymptotic theory of concomitants of order statistics
J. Appl. Prob. 11, 762–770

David, H.A., Groeneveld, R.A. (1982)
Measures of local variation in a distribution: Expected length of spacings and variances of order statistics
Biometrika 69, 227–232

David, H.A., Joshi, P.C. (1968)
Recurrence relations between moments of order statistics for exchangeable variates
Ann. Math. Statist. 39, 272–274

David, H.A., Shu, V.S. (1978)
Robustness of location estimators in the presence of an outlier
in: David, H.A. (ed.), Contributions to Survey Sampling and Applied Statistics, 235–250, Academic Press, New York

David, H.A., O'Connell, M.J., Yang, S.S. (1977)
Distribution and expected value of the rank of a concomitant of an order statistic
Ann. Statist. 5, 216–223

Deheuvels, P. (1983)
The strong approximation of extremal processes II
Z. Wahrscheinlichkeitsth. verw. Geb. 62, 7–15

Deheuvels, P. (1984)
The characterization of distributions by order statistics and record values – a unified approach
J. Appl. Prob. 21, 326–334
Correction, J. Appl. Prob. 22, 997 (1985)

Deheuvels, P., Nevzorov, V.B. (1994)
Limit laws for K–record times
J. Stat. Plan. Inf. 38, 279–307

Deshpande, J.V., Kochar, S.C. (1983)
Dispersive ordering is the same as tail–ordering
Adv. Appl. Prob. 15, 686–687

Doksum, K. (1969)
Starshaped transformations and the power of rank tests
Ann. Math. Statist. 40, 1167–1176

Dykstra, R.L. (1985)
Ordering, Starshaped
in: Kotz, S., Johnson, N.L. (eds.), Encyclopedia of Statistical Sciences, Vol. 6
Wiley, New York

Dziubdziela, W., Kopociński, B. (1976)
Limiting properties of the k–th record values
Applicationes Mathematicae 15, 187–190

Ferguson, T.S. (1964)
A characterization of the exponential distribution
Ann. Math. Statist. 35, 1199–1207

Fisz, M. (1958)
Characterization of some probability distributions
Skand. Aktuarietidskrift 41, 65–67

Gajek, L., Gather, U. (1989)
Moment inequalities for order statistics with applications to characterizations of distributions
Universität Dortmund, Fachbereich Statistik, Forschungsbericht 89/11

Gajek, L., Gather, U. (1991)
Moment inequalities for order statistics with applications to characterizations of distributions
Metrika 38, 357–367

Galambos, J. (1975)
Characterization of probability distributions by properties of order statistics I
in: Patil, G.P. et al. (eds.), Statistical Distributions in Scientific Work, Vol. 3, 71–88
Reidel, Dordrecht

Galambos, J. (1978)
The Asymptotic Theory of Extreme Order Statistics
Wiley, New York

Galambos, J., Kotz, S. (1978)
Characterizations of Probability Distributions
Springer, New York

Gather, U. (1984)
Tests und Schätzungen in Ausreißermodellen
Habilitationsschrift, Aachen University of Technology

Gather, U., Rauhut, B. (1990)
The outlier behaviour of probability distributions
J. Stat. Plan. Inf. 26, 237–252

Glick, N. (1978)
Breaking records and breaking boards
The American Math. Monthly 85, 2–26

Gould, H.W. (1972)
Combinatorial Identities
Morgantown

Govindarajulu, Z. (1963)
On moments of order statistics and quasi–ranges from normal populations
Ann. Math. Statist. 34, 633–651

Gradstein, I., Ryshik, I. (1981)
Summen–, Produkt– und Integraltafeln, Band 1
Harri Deutsch, Thun

Gröbner, W., Hofreiter, N. (1961)
Integraltafel, Erster Teil: Unbestimmte Integrale
Springer, Wien

Gröbner, W., Hofreiter, N. (1961)
Integraltafel, Zweiter Teil: Bestimmte Integrale
Springer, Wien

Grudzień, Z., Szynal, D. (1983)
On the expected values of k–th record values and associated characterizations of distributions
Prob. and Stat. Decision Theory Vol. A, 119–127

Gumbel, E.J. (1954)
The maxima of the mean largest value and of the range
Ann. Math. Statist. 25, 76–84

Gupta, R.C. (1984)
Relationships between order statistics and record values and some characterization results
J. Appl. Prob. 21, 425–430

Gupta, R.C., Kirmani, S.N.U.A. (1987)
On order relations between reliability measures
Commun. Statist. – Stochastic Models 3, 149–156

Gupta, R.C., Kirmani, S.N.U.A. (1988)
Closure and monotonicity properties of nonhomogeneous Poisson processes and record values
Probability in the Engineering and Informational Sciences 2, 475–484

Gupta, R.C., Kirmani, S.N.U.A. (1990)
The role of weighted distributions in stochastic modeling
Commun. Statist. – Theory Meth. 19, 3147–3162

Gupta, S.S., Shah, B.K. (1965)
Exact moments and percentage points of the order statistics and the distribution of the range from the logistic distribution
Ann. Math. Statist. 36, 907–920

Harter, H.L. (1988)
History and role of order statistics
Commun. Statist. – Theory Meth. 17, 2091–2107

Hartley, H.O., David, H.A. (1954)
Universal bounds for mean range and extreme observation
Ann. Math. Statist. 25, 85–99

Hoeffding, W. (1953)
On the distribution of the expected values of the order statistics
Ann. Math. Statist. 24, 93–100

Houchens, R.L. (1984)
Record value theory and inference
Dissertation, University of California, Riverside

Huang, J.S. (1974a)
On a theorem of Ahsanullah and Rahman
J. Appl. Prob. 11, 216–218

Huang, J.S. (1974b)
Characterizations of the exponential distribution by order statistics
J. Appl. Prob. 11, 605–608

Huang, J.S. (1975)
Characterization of distributions by the expected values of the order statistics
Ann. Inst. Statist. Math. 27, 87–93

Huang, J.S. (1989)
Moment problem of order statistics: A review
International Statistical Review 57, 59–66

Huang, J.S., Hwang, J.S. (1975)
L_1–completeness of a class of beta densities
in: Patil, G.P. et al. (eds.), Statistical Distributions in Scientific Work, Vol. 3, 137–141, Reidel, Dordrecht

Huber, P.J. (1981)
Robust Statistics
Wiley, New York

Hwang, J.S. (1978)
A note on Bernstein and Müntz–Szasz theorems with applications to the order statistics
Ann. Inst. Statist. Math. 30 A, 167–176

Hwang, J.S. (1983)
On a generalized moment problem
Proc. American Math. Soc. 87, 88–89

Hwang, J.S., Lin, G.D. (1984a)
Characterizations of distributions by linear combinations of moments of order statistics
Bulletin of the Institute of Mathematics, Academia Sinica 12, 179–202

Hwang, J.S., Lin, G.D. (1984b)
On a generalized moment problem II
Proc. American Math. Soc. 91, 577–580

Hwang, J.S., Lin, G.D. (1987)
Nonlinear approximation and moment problem
in: Rassias, Th.M. (ed.), Nonlinear Analysis, 337–353
World Scientific Publ., Singapore

Johnson, N.L., Kotz, S. (1970)
Distributions in Statistics, Continuous Univariate Distributions 1
Houghton Mifflin, Boston

Johnson, N.L., Kotz, S. (1972)
Distributions in Statistics, Continuous Multivariate Distributions
Wiley, New York

Joshi, P.C. (1977)
Recurrence relations between moments of order statistics from exponential and truncated exponential distributions
Sankhyā 39 B, 362–371

Joshi, P.C., Balakrishnan, N. (1982)
Recurrence relations and identities for the product moments of order statistics
Sankhyā 44 B, 39–49

Kadane, J.B. (1974)
A characterization of triangular arrays which are expectations of order statistics
J. Appl. Prob. 11, 413–416

Kagan, A.M., Linnik, I.V., Rao, C.R. (1973)
Characterization Problems in Mathematical Statistics
Wiley, New York

Kakosyan, A., Klebanov, L.B., Melamed, J.A. (1984)
Characterization of Distributions by the Method of Intensively Monotone Operators
Springer, Berlin

Kamps, U. (1990a)
A characterizing property of Cauchy, doubly truncated Cauchy and related distributions
Technical Report, Aachen University of Technology

Kamps, U. (1990b)
Characterizations of the exponential distribution by weighted sums of iid random variables
Statistical Papers 31, 233–237

Kamps, U. (1991a)
Inequalities for moments of order statistics and characterizations of distributions
J. Stat. Plan. Inf. 27, 397–404

Kamps, U. (1991b)
A general recurrence relation for moments of order statistics in a class of probability distributions and characterizations
Metrika 38, 215–225

Kamps, U. (1992a)
Identities for the difference of moments of successive order statistics and record values
Metron 50, 179–187

Kamps, U. (1992b)
Characterizations of the exponential distribution by equality of moments
Allgemeines Statistisches Archiv 76, 122–127

Kamps, U., Mattner, L. (1993)
An identity for expectations of functions of order statistics
Metrika 40, 361–365

Khan, A.H., Khan, I.A. (1987)
Moments of order statistics from Burr distribution and its characterizations
Metron 45, 21–29

Khan, A.H., Khan, R.U. (1983)
Recurrence relations between moments of order statistics from generalized gamma distribution
Journal of Statistical Research 17, 75–82

Khan, A.H., Yaqub, M., Parvez, S. (1983)
Recurrence relations between moments of order statistics
Naval Research Logistics Quarterly 30, 419–441
Corrigendum 32, 693 (1985)

Kim, J.S., Proschan, F., Sethuraman, J. (1988)
Stochastic comparisons of order statistics, with applications in reliability
Commun. Statist. – Theory Meth. 17, 2151–2172

Kirmani, S.N.U.A., Beg, M.I. (1984)
On characterization of distributions by expected records
Sankhyā 46 A, 463–465

Kochar, S.C. (1990)
Some partial ordering results on record values
Commun. Statist. – Theory Meth. 19, 299–306

Konheim, A.G. (1971)
A note on order statistics
American Math. Monthly 78, 524

Lawless, J.F. (1982)
Statistical Models and Methods for Lifetime Data
Wiley, New York

Leadbetter, M.R., Lindgren, G., Rootzén, H. (1983)
Extremes and Related Properties of Random Sequences and Processes
Springer, New York

Lewis, T., Thompson, J.W. (1981)
Dispersive distributions, and the connection between dispersivity and strong unimodality
J. Appl. Prob. 18, 76–90

Lieblein, J. (1955)
On moments of order statistics from the Weibull distribution
Ann. Math. Statist. 26, 330–333

Lin, G.D. (1984)
A note on equal distributions
Ann. Inst. Statist. Math. 36A, 451–453

Lin, G.D. (1986)
On a moment problem
Tôhoku Math. Journ. 38, 595–598

Lin, G.D. (1987)
On characterizations of distributions via moments of record values
Probab. Th. Rel. Fields 74, 479–483

Lin, G.D. (1988a)
Characterizations of uniform distributions and of exponential distributions
Sankhyā 50 A, 64–69

Lin, G.D. (1988b)
Characterizations of distributions via relationships between two moments of order statistics
J. Stat. Plan. Inf. 19, 73–80

Lin, G.D. (1989a)
Characterizations of distributions via moments of order statistics: A survey and comparison of methods
in: Dodge, Y. (ed.), Statistical Data Analysis and Inference, 297–307
North–Holland, Amsterdam

Lin, G.D. (1989b)
The product moments of order statistics with applications to characterizations of distributions
J. Stat. Plan. Inf. 21, 395–406

Lin, G.D., Huang, J.S. (1987)
A note on the sequence of expectations of maxima and of record values
Sankhyā 49 A, 272–273

Malik, H.J. (1966)
Exact moments of order statistics from the Pareto distribution
Skand. Aktuarietidskrift 49, 144–157

Malik, H.J. (1967)
Exact moments of order statistics from a power–function distribution
Skand. Aktuarietidskrift 50, 64–69

Malik, H.J., Balakrishnan, N., Ahmed, S.E. (1988)
Recurrence relations and identities for moments of order statistics, I: Arbitrary continuous distribution
Commun. Statist. – Theory Meth. 17, 2623–2655

Mallows, C.L. (1973)
Bounds on distribution functions in terms of expectations of order statistics
Ann. Probab. 1, 297–303

Mann, N.R., Schafer, R.E., Singpurwalla, N.D. (1974)
Methods for Statistical Analysis of Reliability and Life Data
Wiley, New York

McCool, J.I. (1982)
Censored Data
in: Kotz, S., Johnson, N.L. (eds.), Encyclopedia of Statistical Sciences, Vol. 1
Wiley, New York

Melnick, E.L. (1964)
Moments of ranked Poisson variates
M.S. Thesis, Virginia Polytechnic Institute

Mitrinović, D.S. (1970)
Analytic Inequalities
Springer, Berlin

Moriguti, S. (1951)
Extremal property of extreme value distributions
Ann. Math. Statist. 22, 523–536

Müntz, C.H. (1914)
Über den Approximationssatz von Weierstraß
Festschrift für H.A. Schwarz, Berlin, 303–312

Nagaraja, H.N. (1978)
On the expected values of record values
Austral. J. Statist. 20, 176–182

Nagaraja, H.N. (1982)
On the non–Markovian structure of discrete order statistics
J. Stat. Plan. Inf. 7, 29–33

Nagaraja, H.N. (1988)
Record values and related statistics – A review
Commun. Statist. – Theory Meth. 17, 2223–2238

Nagaraja, H.N. (1990)
Some reliability properties of order statistics
Commun. Statist. – Theory Meth. 19, 307–316

Natanson, I.P. (1961)
Theorie der Funktionen einer reellen Veränderlichen
Akademie–Verlag, Berlin

Nayak, S.S. (1981)
Characterizations based on record values
J. Indian Statist. Ass. 19, 123–127

Nelson, W. (1982)
Applied Life Data Analysis
Wiley, New York

Nevzorov, V.B. (1986a)
On the k–th record times and their generalizations
Zap. Nauchn. Sem. Leningrad. Otdel. Mat. Inst. Steklov 153, 115–121

Nevzorov, V.B. (1986b)
Record times and their generalizations
Theory Probab. Appl. 31, 554–556

Nevzorov, V.B. (1987)
Records
Theory Probab. Appl. 32, 201–228

Oja, H. (1985)
Ordering of Distributions, Partial
in: Kotz, S., Johnson, N.L. (eds.), Encyclopedia of Statistical Sciences, Vol. 6
Wiley, New York

Papathanasiou, V. (1990)
Some characterizations of distributions based on order statistics
Statistics & Probability Letters 9, 145–147

Patel, J.K. (1983)
Hazard rate and other classifications of distributions
in: Kotz, S., Johnson, N.L. (eds.), Encyclopedia of Statistical Sciences, Vol. 3
Wiley, New York

Patwardhan, G. (1988)
Tests for exponentiality
Commun. Statist. – Theory Meth. 17, 3705–3722

Pfeifer, D. (1979)
'Record Values' in einem stochastischen Modell mit nicht–identischen Verteilungen
Dissertation, Aachen University of Technology

Pfeifer, D. (1982a)
Characterizations of exponential distributions by independent non–stationary record increments
J. Appl. Prob. 19, 127–135
Correction 19, 906

Pfeifer, D. (1982b)
The structure of elementary pure birth processes
J. Appl. Prob. 19, 664–667

Pfeifer, D. (1989)
Einführung in die Extremwertstatistik
Teubner, Stuttgart

Plackett, R.L. (1947)
Limit of the ratio of mean range to standard deviation
Biometrika 34, 120–122

Pollak, M. (1973)
On equal distributions
Ann. Statist. 1, 180–182

Pólya, G., Szegö, G. (1964)
Aufgaben und Lehrsätze aus der Analysis, Erster Band
Springer, Berlin

Pyke, R. (1965)
Spacings
J. Royal Statist. Soc. B 27, 395–436

Rahman, N.A. (1964)
Some generalisations of the distributions of product statistics arising from rectangular populations
J. Am. Statist. Ass. 59, 557–563

Reiss, R.D. (1989)
Approximate Distributions of Order Statistics
Springer, New York

Rényi, A. (1953)
On the theory of order statistics
Acta Math., Acad. Sci. Hungar. 4, 191–231

Resnick, S.I. (1973a)
Limit laws for record values
Stoch. Proc. Appl. 1, 67–82

Resnick, S.I. (1973b)
Record values and maxima
Ann. Probab. 1, 650–662

Resnick, S.I. (1987)
Extreme Values, Regular Variation, and Point Processes
Springer, New York

Rider, P.R. (1955)
The distribution of the product of maximum values in samples from a rectangular distribution
J. Am. Statist. Ass. 50, 1142–1143

Rodriguez, R.N. (1982)
Burr distributions
in: Kotz, S., Johnson, N.L. (eds.), Encyclopedia of Statistical Sciences, Vol. 1
Wiley, New York

Rohatgi, V.K. (1976)
An Introduction to Probability Theory and Mathematical Statistics
Wiley, New York

Rohatgi, V.K., Saleh, A.K.MD.E. (1988)
A class of distributions connected to order statistics with nonintegral sample size
Commun. Statist. – Theory Meth. 17, 2005–2012

Ross, S.M. (1983)
Stochastic Processes
Wiley, New York

Rossberg, H.J. (1960)
Über die Verteilungsfunktionen der Differenzen und Quotienten von Ranggrößen
Math. Nachrichten 21, 37–79

Rüschendorf, L. (1985)
Two remarks on order statistics
J. Stat. Plan. Inf. 11, 71–74

Saleh, A.K.MD.E. (1976)
Characterization of distributions using expected spacings between consecutive order statistics
Journal of Statistical Research 10, 1–13

Sathe, Y.S., Dixit, U.J. (1990)
On a recurrence relation for order statistics
Statistics & Probability Letters 9, 1–4

Sen, P.K. (1959)
On the moments of the sample quantiles
Calcutta Statistical Association Bulletin 9, 1–19

Shah, B.K. (1970)
Note on moments of a logistic order statistic
Ann. Math. Statist. 41, 2150–2152

Shaked, M. (1982)
Dispersive ordering of distributions
J. Appl. Prob. 19, 310–320

Shaked, M. (1985)
Ordering Distributions by Dispersion
in: Kotz, S., Johnson, N.L. (eds.), Encyclopedia of Statistical Sciences, Vol. 6
Wiley, New York

Shorrock, R.W. (1972)
A limit theorem for inter–record times
J. Appl. Prob. 9, 219–223

Stigler, S.M. (1977)
Fractional order statistics, with applications
J. Am. Statist. Ass. 72, 544–550

Sukhatme, P.V. (1937)
Tests of significance for samples of the χ^2 population with two degrees of freedom
Ann. Eugenics 8, 52–56

Szász, O. (1916)
Über die Approximation stetiger Funktionen durch lineare Aggregate von Potenzen
Math. Ann. 77, 482–496

Tadikamalla, P.R. (1980)
A look at the Burr and related distributions
International Statistical Review 48, 337–344

Takahasi, K. (1988)
A note on hazard rates of order statistics
Commun. Statist.–Theory Meth. 17, 4133–4136

Tata, M.N. (1969)
On outstanding values in a sequence of random variables
Z. Wahrscheinlichkeitsth. verw. Geb. 12, 9–20

Terrell, G.R. (1983)
A characterization of rectangular distributions
Ann. Probab. 11, 823–826

Tiku, M.L. (1988)
Order statistics in goodness–of–fit tests
Commun. Statist. – Theory Meth. 17, 2369–2387

Too, Y.H., Lin, G.D. (1989)
Characterizations of uniform and exponential distributions
Statistics & Probability Letters 7, 357–359

Whitt, W. (1988)
Stochastic Ordering
in: Kotz, S., Johnson, N.L. (eds.), Encyclopedia of Statistical Sciences, Vol. 8
Wiley, New York

Witte, H.J. (1988)
Some characterizations of distributions based on the integrated Cauchy functional equation
Sankhyā 50 A, 59–63

Witting, H. (1985)
Mathematische Statistik I
Teubner, Stuttgart

Yang, S.S. (1977)
General distribution theory of the concomitants of order statistics
Ann. Statist. 5, 996–1002

Young, D.H. (1967)
Recurrence relations between the P.D.F.'s of order statistics of dependent variables, and some applications
Biometrika 54, 283–292

Author Index

Subject Index